READING ABOUT SCIENCE
Skills and Concepts

BOOK D

John Mongillo
Beth Atwood
Kevin M. Carr
Linda J. Carr
Claudia Cornett
Jackie Harris
Josepha Sherman
Vivian Zwaik

Phoenix Learning Resources

ISBN 0-7915-2204-0

2 3 4 5 6 7 8 9 0 05

Authors

John Mongillo, Senior Author and General Editor
Science Writer and Editor
Saunderstown, Rhode Island

Beth S. Atwood
Writer and Reading Consultant
Durham, Connecticut

Kevin M. Carr
Teacher and Writer
Honolulu, Hawaii

Linda J. Carr
Writer and Psychologist
Honolulu, Hawaii

Claudia Cornett
Professor Emerita
Wittenberg University

Jackie Harris
Medical and Science Editor
Wethersfield, Connecticut

Josepha Sherman
Writer and Science Editor
Riverdale, New York

Vivian Zwaik
Writer and Educational Consultant
Wayne, New Jersey

CONTENTS

Do you enjoy the world around you? Do you ever wonder why clouds have so many different shapes and what keeps planes up in the air? Did you ever want to explore a cave or find out why volcanoes erupt or why the earth shakes? If you can answer yes to any of these questions, then you will enjoy reading about science.

The world of science is a world of observing, exploring, predicting, reading, experimenting, testing, and recording. It is a world of trying and failing and trying again until, at last, you succeed. In the world of science, there is always some exciting discovery to be made and something new to explore.

Four Areas of Science

READING ABOUT SCIENCE explores four areas of science: life science, earth-space science, physical science, and environmental science. Each book in this series contains a unit on each of the four areas.

Life science is the study of living things. Life scientists explore the world of plants, animals, and humans. Their goal is to find out how living things depend upon each other for sur-

vival and to observe how they live and interact in their environments, or surroundings.

Life science includes many specialized areas, such as botany, zoology, and ecology. *Botanists* work mainly with plants. *Zoologists* work mostly with animals. *Ecologists* are scientists who study the effects of air pollution, water pollution, and noise pollution on living things.

Earth-space science is the study of our Earth and other bodies in the solar system. Some earth-space scientists are *meteorologists*, who study climate and weather; *geologists*, who study the earth, the way it was formed, and its makeup, rocks and fossils, earthquakes, and volcanoes; *oceanographers*, who study currents, waves, and life in the oceans of the world; and *astronomers*, who study the solar system, including the sun and other stars, moons, and planets.

Physical science is the study of matter and energy. *Physicists* are physical scientists who explore topics such as matter, atoms, and nuclear energy. Other physical scientists study sound, magnetism, heat, light, electricity, water, and air. *Chemists* develop the substances used in medicine, clothing, food, and many other things.

Environmental science is the study of the forces and conditions that surround and influence all living and nonliving things. Environmental science involves all of the other sciences-life, earth-space, and physical.

If you want to know more about one or more of these areas of science, check the bibliography at the back of this book for suggested additional readings.

Steps to Follow

The suggestions that follow will help you use this book:

A. Study the photo or drawing that goes with the story. Read the title and the sentences that are printed in the sidebar next to each story. These are all clues to what the story is about.

B. Study the words for the story in the list of Words to Know at the beginning of each unit. You will find it easier to read the story if you understand the meanings of these words. Many times, you will find the meaning of the word right in the story.

When reading the story, look for clues to important words or ideas. Sometimes words or phrases are underlined. Pay special attention to these clues.

C. Read the story carefully. Think about what you are reading. Are any of the ideas in the story things that you have heard or read about before?

D. As you read, ask yourself questions. For example, "Why did the electricity go off?" "What caused the bears to turn green?" Many times, your questions are answered later in the story. Questioning helps you to understand what the author is saying. Asking questions also gets you ready for what comes next in the story.

E. Pay special attention to diagrams, charts, and other visual aids. They will often help you to understand the story better.

F. After you read the story slowly and carefully, you are ready to answer the questions on the questions page. If the book you have is part of a classroom set, you should write your answers in a special notebook or on paper that you can keep in a folder. Do not write in this book without your teacher's permission.

Put your name, the title of the story, and its page number on a sheet of paper. Read each question carefully. Record the question number and your answer on your answer paper.

The questions in this book check for the following kinds of comprehension, or understanding:

1. *Science vocabulary comprehension.* This kind of question asks you to remember the meaning of a word or term used in the story.

2. *Literal comprehension*. This kind of question asks you to remember certain facts that are given in the story. For example, the story might state that a snake was over 5 feet long. A literal question would ask you: "How long was the snake?"

3. *Interpretive comprehension*. This kind of question asks you to think about the story. To answer the question, you must decide what the author means, not what is said, or stated, in the story. For example, you may be asked what the main idea of the story is, what happened first, or what caused something to happen in the story.

4. *Applied comprehension*. This kind of question asks you to use what you have read to (1) solve a new problem, (2) interpret a chart or graph, or (3) put a certain topic under its correct heading, or category.

You should read each question carefully. You may go back to the story to help you find the answer. The questions are meant to help you learn how to read more carefully.

G. When you complete the questions page, turn it in to your teacher. Or, with your teacher's permission, check your answers against the answer key in the Teacher's Guide. If you made a mistake, find out what you did wrong. Practice answering that kind of question, and you will do better the next time.

H. Turn to the directions that tell you how to keep your Progress Charts. If you are not supposed to write in this book, you may make a copy of each chart to keep in your READING ABOUT SCIENCE folder or notebook. You may be surprised to see how well you can read science.

PRONUNCIATION GUIDE

Some words in this book may be unfamiliar to you and difficult for you to pronounce. These words are printed in italics. Then they are spelled according to the way they are said,or pronounced. This phonetic spelling appears in parentheses next to the words. The pronunciation guide below will help you say the words.

ă	pat	î	dear, deer, fierce,	p	pop	zh	garage, pleasure;
ā	aid, fey, pay		mere	r	roar		vision
â	air, care, wear	j	judge	s	miss, sauce, see	ə	about, silent
ä	father	k	cat, kick, pique	sh	dish, ship		pencil, lemon,
b	bib	l	lid, needle	t	tight		circus
ch	church	m	am, man, mum	th	path, thin	ər	butter
d	deed	n	no, sudden	*th*	bathe, this		
ĕ	pet, pleasure	ng	thing	ŭ	cut, rough		
ē	be, bee, easy,	ŏ	horrible, pot	û	circle, firm, heard,		
	leisure	ō	go, hoarse, row,		term, turn, urge,		
f	fast, fife, off,		toe		word		STRESS
	phase, rough	ô	alter, caught, for,	v	cave, valve, vine		Primary stress '
g	gag		paw	w	with		bi·ol'o·gy
h	hat	oi	boy, noise, oil	y	yes		[bī ŏl'ejē]
hw	which	ou	cow, out	yoō	abuse, use		Secondary stress'
ĭ	pit	o͝o	took	z	rose, size,		bi'o·log'i·cal
ī	by, guy, pie	o͞o	boot, fruit		xylophone, zebra		[bī'elŏj'īkel]

The key to pronunciation above is reprinted by permission from *The American Heritage School Dictionary* copyright © 1977, by Houghton Mifflin Company

LIFE SCIENCE

All birds have feathers. Their feathers serve many purposes. Brightly colored birds such as the peacock show off their beautiful feathers to attract a mate. Sometimes color helps a bird hide from enemies by making it blend in with the surroundings. Feathers make a bird's body smooth so that it glides through the air easily. Feathers also keep a bird warm as it flies.

WORDS TO KNOW

The Praying Mantis
species, a group of highly similar plants or animals

predator, an animal that lives by feeding on other animals

cocoon, a silky case that certain insects spin about themselves for shelter

Lana Learns to Talk
symbols, a mark that stands for, represents, something else

communicate, to give or exchange information

Tool Users
pound, to hit with force

termite, an antlike insect that lives in a colony and is very destructive to wooden structures

shellfish, any animal that lives in water and has a shell

The Largest Owl in North America
talons, the claws of birds that hunt animals

prey, an animal hunted for food by another animal

swoop, to sweep down upon

The Horseshoe Crab—a Living Fossil
fossil, hardened remains of plant or animal life

hinged, having a hinge. A hinge is a joint on which something swings, such as a door or gate, etc.

spiked, comes to a point

pincers, claws for holding on to something

On the Trail of the Grizzly Bear
extinct, no longer existing

inject, to force a fluid into a body

Animals and Earthquakes
behavior, the way something acts

fault, a break, or fracture, in a rock formation

energy, force or power

onset, beginning

accurately, exactly, free from errors

The Octopus—a Sea Creature
mollusks, a large group of sea animals without backbones, with a soft body often enclosed in a hard shell

mussels, freshwater mollusks

protective, offering shield from injury and danger

No Longer a Threat
branded, called, labeled

bounty, a reward

endangered, an animal or plant in danger of no longer existing

Why Muscles Move

tissue, a group of similar cells in a body or organ that perform the same function

contracting, reducing in size (length)

voluntary, made to act by one's own free choice, intentional

involuntary, not consciously controlled

Studying the Lunch Bunch

nutritionist, a person who specializes in the study of nutrition.

nutrition, the steps by which living organisms take in food and use it to keep healthy and grow.

satisfy, to fulfill a need or desire

calories, a calorie is the amount of heat it takes to raise one kilogram of water (2.2 pounds) of water 1 degree centigrade; used as a measure of energy available in foods.

dietitians, an expert in the kinds and amounts of food necessary for health

nourish, to feed or sustain life and growth

A Home for the Birds

pollution, dirtied, ruined

An Amazing Little Weed

perennial, yearly, happening every year

notched, having a V-shaped edge or surface

The Desert Food Chain

survive, to stay alive

glucose, a sugar occurring naturally in plants

nutrients, foods that feed the plant

Fungi, a Different Kind of Organism

environment, surroundings

various, different from one another

organic, coming from living things

Your Incredible Breathing Machine

inhale, to breathe in, take air in

exhale, to breathe out

exchange, to replace one thing with another

respiratory, breathing

The Praying Mantis

This insect is helpful to humans.

The praying mantis is quite a sight to see. While it is resting or when it is getting ready to attack its prey, this mantis folds its front legs as if it were praying.

There are about 1,700 known species, or kinds, of mantises. Most mantises are green, brown, or gray. But certain species are brightly colored. The mantis is a predator (prĕd′ə tər), an animal that lives by feeding on other animal life. This predator eats all sorts of insects, including harmful garden insects that chew up plants.

The female mantis lays her eggs, one at a time, in a cocoon (kə ko͞on′). The cocoon is made of a foam-like material that flows from the female's body. It hardens into different shapes. The female produces several cocoons and always attaches them to something, like a stone, a twig, or a leaf. A single cocoon may hold 50 to 400 eggs!

The mantis's eggs spend the winter in the cocoon and hatch in the spring. Then it takes the mantis about 3 to 5 months to develop fully. As adults, these predatory insects live only for a few months.

The mantis is often invisible both to its enemies and to its prey. Some praying mantises look like green or brown leaves, and others look like flowers. All in all, the mantis is one of the animal world's most interesting insects.

1. The word in the story that describes an animal that feeds on other animals is _____ .

2. The female mantis usually produces
   a. less than 50 eggs.
   b. at least 400 cocoons.
   c. more than one cocoon.

3. What does the praying mantis do that makes it look as if it were praying?
   a. It folds its front legs.
   b. It changes color from brown to green.
   c. It makes cocoons for its eggs.

4. The story says that the praying mantis is often invisible to its enemies and its prey. How is this possible?
   a. It is able to disappear completely from view.
   b. It may be mistaken for a leaf that is similar in shape and color.
   c. It is small enough to hide under stones when an enemy is nearby.

5. The praying mantis is fully developed when it hatches.
   a. True
   b. False
   c. The story does not say.

6. Suppose you found a praying mantis in your garden. What would be the best thing to do?
   a. Chase it away before it eats the plants.
   b. Ignore it and let it stay in the garden.
   c. Kill it, because it will harm other insects.

Lana Learns to Talk

Lana is smart. She asks for food, and she even asks for music. What is strange about that?

Lana is a chimpanzee that uses a computer (kəm pyo͞o′tər) to "talk." To do this, Lana must use a special language made up of pictures and symbols that stand for words. If Lana wants a banana, she simply pushes a button on the computer. The correct button has a picture that stands for banana. So if Lana pushes that button, the computer gives her a banana.

Lana can even "write" sentences. The sentences may not contain all the words that a human would use when talking. But each sentence must make sense in order for Lana to be understood. Suppose Lana wants milk and she punches out *Please, machine, make milk*. Lana will not get the milk. She must first correct the sentence. When she punches out *Please, machine, give milk*, Lana gets her milk.

Lana's trainer is also her friend. They communicate (kə myo͞o′nə kāt′), or exchange information, with each

other by using the computer. Lana started to use the computer to communicate when she was only 3 years old. She won't be fully grown until she is between 12 and 16 years old. So there is still a lot more Lana can learn. Now that she knows something about letters, scientists are testing her ability with numbers. What scientists are learning from Lana is being used to help people with language problems.

1.   The word in the story that means to "exchange information" is _____.

2.   To talk to her trainer, Lana must use a machine called a _____.

3.   Lana's sentences are made up of all the words that humans would use when talking.

    a.   True

    b.   False

    c.   The story does not say.

4.   What might happen because of what the research scientists are doing with chimps like Lana?

    a.   Humans who have trouble communicating are being helped.

    b.   Chimps may talk using complete sentences.

    c.   Humans will learn to speak to chimps without computers.

5.   Before Lana could use the computer, it was necessary for her to

    a.   learn to identify certain pictures and symbols.

    b.   read and write many words and sentences.

    c.   learn how to talk with scientists.

6.   If Lana wanted to play a game of catch with her friend, which sentence would she punch into the computer?

    a.   Please, friend, play catch with Lana.

    b.   Lana wants machine play catch.

    c.   Friend wants play game, Lana.

Tool Users

When you want to pound a nail into a board, you use a hammer. When you want to eat soup from a bowl, you use a spoon. Both hammers and spoons are tools.

Some animals also use tools. The *shrike* (shrīk) is a bird that uses a tool to catch its *prey* (prā). *Prey* is the word used to describe any animal that is hunted and caught for food. The shrike's tools are the sharp, pointed thorns found on some trees and bushes. First, the shrike catches its meal, usually an insect, a mouse, or a small bird. Then, after sticking its prey onto a thorn, the shrike uses its sharp bill, or beak, to tear its prey into small, bite-sized bits.

In Africa, chimpanzees use grass stems or tree twigs as tools. The chimp pokes the stems or twigs into termite holes. When it feels a termite biting on the end of the twig, the chimp pulls the twig up. If the termite has not let go of the twig, the chimp has a tasty insect to eat!

The California sea otter feeds on shellfish. When the otter is hungry, it dives to the bottom of the sea. Soon, it comes back up to the surface with a shellfish and a rock. Floating on its back, the sea otter places the shellfish on its chest. Then, holding the rock in one paw, the sea otter smashes open the shell and eats the fish inside.

1. *Prey* is a term used to describe any animal that

    a. has sharp, pointed thorns.

    b. is used as a tool.

    c. is hunted and caught for food.

2. A shrike uses a _____ as a tool.

3. According to the story, a termite makes a tasty meal for chimps.

    a. True

    b. False

    c. The story does not say.

4. Just after returning to the surface of the sea with its food, the sea otter will

    a. place the shellfish on a rock.

    b. turn over on its back and float.

    c. use its prey as a tool.

5. What is the main idea of this story?

    a. Animals, as well as people, use tools.

    b. Termites need to eat wood to stay alive.

    c. Shellfish are found in California waters.

6. If you saw a chimpanzee pulling on a twig, you would probably think it was busy

    a. building its home.

    b. hiding from its enemies.

    c. catching its food.

The Largest Owl in North America

You are camping in the forest late one moonless night. Suddenly, you hear a horrible screech, then a deep call: "HOO, HOO, HOO!"

What you hear is probably the sound of the great horned owl, also called the "hoot owl." The great horned owl is the largest owl in North America. The giant owl is as long as a large house cat and has a wingspread of nearly 5 feet. It is so big that it can attack skunks and rabbits as well as smaller animals.

The great horned owl is yellow and brown with white and black markings. It has a white throat and deep yellow eyes. It also has sharp claws called *talons* (tăl′ənz). It uses its talons to kill its prey, or food. Good sight and a keen sense of smell help the hoot owl swoop down on anything that moves.

An owl's eyes are so good that it can see well even in dim light. This makes it an able night hunter. An owl cannot, however, move its eyes from side to side the way humans can. Instead, it must move its whole head when it wants to see something. In fact, an owl can turn its head almost completely around.

The great horned owl also has excellent hearing. The feathers around its ears make it look like it has horns. The feathers, however, form a cup that catches sounds from many directions. An animal can be as quiet as a mouse, but that won't keep the owl from hearing it move!

QUESTIONS

1. An owl's sharp claws are called

    a. horns.

    b. swoops.

    c. talons.

2. Another name for the great horned owl is the _____.

3. An owl can see well in dim light.

    a. True

    b. False

    c. The story does not say.

4. A great horned owl probably eats

    a. mice, squirrels, and rabbits.

    b. cows, pigs, and chickens.

    c. tigers, lions, and horses.

5. Great horned owls seem scary because they

    a. suck the blood of animals.

    b. have long fangs.

    c. screech horribly.

6. If your eyes were like those of a great horned owl, then you would

    a. not have any eyelids.

    b. not be able to sleep at night.

    c. have to turn your head to look around.

The Horseshoe Crab—A Living Fossil

The horseshoe crab is a living fossil.

A sea animal called the horseshoe crab is found buried along sandy ocean shores. But it really isn't related to any other living thing in the oceans. In fact, its closest relative, the spider, is found on land.

Most crabs have shells that are hard, and pieces can be broken off easily. But the horseshoe crab has a tough shell that is in two hinged parts. Spiders also have two body parts. The horseshoe crab has a long spiked tail called a *telson* (tĕl'sən). It looks dangerous, but it isn't. If the animal gets turned over on its back, it uses its tail to turn itself right side up again.

The horseshoe crab has six pairs of legs. The small pair has *pincers* (pĭn'sərz). As the animal moves its small legs, the pincers grip and crush the horseshoe crab's food. Horseshoe crabs feed on clam worms, soft-shelled clams, and other small animals. The other pairs of legs are used for walking and feeling.

Horseshoe crabs have been on Earth for about 250 million years and are considered "living *fossils*" (fŏs'əlz).

Fossils are the remains of ancient plant and animal life. Fossils have been found of horseshoe crab relatives that lived when dinosaurs were alive. The fossils were found in rocks that are 100 million years old. Yet the fossil relatives look just the same as our present-day horseshoe crabs.

1. The remains, often found in rock, of an animal or a plant that lived long ago are called _____.

2. The shell of the horseshoe crab is hard and breaks off easily.

    a. True

    b. False

    c. The story does not say.

3. According to the story, the horseshoe crab might use its tail to

    a. turn itself right side up.

    b. push food into its mouth.

    c. bury itself along the sandy shore.

4. Horseshoe crabs are called living fossils because they

    a. live on and on in the sea forever.

    b. look unchanged from their ancient relatives.

    c. feed off the remains of other animal fossils.

5. In order to crush its food, the horseshoe crab must move

    a. its small legs.

    b. all six legs.

    c. its spiked tail.

6. Under which of the following groups would you list the horseshoe crab?

    a. Sea Animals with Tough Shells

    b. Sea Animals with Hinged Body Parts

    c. Sea Animals with Dangerous Tails

On the Trail of the Grizzly Bear

The grizzly bear is in danger.

The fierce grizzly bear was a danger to the lives of early settlers in the mountains of the western United States. This huge animal stole food. It also took the lives of many people, cows, and horses. For these reasons, the settlers went to war against the grizzly. By 1931 the grizzly bear had been killed off in five western states. Today, people are worried that the grizzly bear may become extinct (ĭk stĭngkt'). That is, the bear may disappear completely from Earth.

To make sure that these bears do not become extinct, scientists need to learn more about the grizzly. To do this, they track, or follow, the bear's trail.

First, the scientists trap the grizzly and inject the bear with a special medicine to make it sleepy. While the bear is asleep, a special collar is put around its neck. The collar contains a kind of radio that makes a beeping noise. The scientists have a radio receiver that picks up the beeps.

In this way, when the bear wakes up and returns to the woods, scientists can trail it safely from a plane or a truck. This method of tracking the bear has helped scientists learn a lot about how the grizzly lives. This knowledge is helping to make sure the grizzly has what it needs to live. Today, some grizzlies live in national parks in Alaska, Canada, the northern Rocky Mountains, and in Yellowstone National Park, Wyoming.

1. When an animal disappears completely from the Earth, that animal is said to be _____.

2. Many grizzly bears were killed by _____.

3. The special bear collar helps scientists learn a lot about the grizzly.

   a. True

   b. False

   c. The story does not say.

4. In order to study a grizzly, which of the following things does a scientist do *first*?

   a. trap the bear

   b. put a collar on the bear

   c. give the bear an injection

5. What is the *main idea* of the story?

   a. We must learn how the grizzly lives to save it from extinction.

   b. Using a radio collar can help scientists track wild animals.

   c. By 1931 the grizzly was extinct in five states.

6. If you wanted to learn more about the grizzly, under which of the following headings would you look?

   a. How to Follow Animal Trails

   b. Mountain Animals of North America

   c. Animals That Are Now Extinct

Animals and Earthquakes

Can animals predict an earthquake?

Are animals able to sense an earthquake *before* it happens? In China, information has been collected about *animal behavior* (bĭ hāv′yər)—that is, how animals act—before earthquakes. Before one earthquake, a wild rabbit twice jumped onto the roof of a farmhouse and would not come down. Dairy cows in their sheds tried to break loose and run outside. Other animals also did strange things.

American scientists are also trying to find out more about earthquakes and animal behavior. Various animals are being tested at places such as the San Andreas fault (fôlt) in California. A fault is a break, or crack, in a large rock formation. The rock on one side of the break moves past the rock on the other side. This sudden movement releases a great deal of energy. And it causes the earth to *quake* (kwāk), or shake.

At one place along the San Andreas fault, chimpanzees were studied. Scientists found that the chimps became very restless before an earthquake. At other locations along the fault, the behavior of kangaroo rats and pocket mice is being studied.

New scientific instruments allow scientists to record how animals behave before, during, and after each earthquake. By studying these changes in animal behavior scientists hope they may someday be able to predict future earthquakes more accurately.

1. This story tells about, scientists who are interested in changes in animal _____ before, during, and after an earthquake.

2. Studies on animal behavior and earthquakes are being done along the _____ in California.

3. An earthquake results when rock along a fault line suddenly moves and releases a lot of _____.

4. Scientists feel that there is probably a connection between changes that occur in the earth before an earthquake and _____ along fault lines.

   a. changes in rock formation

   b. unusual behavior in animals

   c. the disappearance of animals

5. The goal of the research described in this story is to

   a. prevent earthquakes from occurring.

   b. save the lives of many people.

   c. keep animals away from earthquakes.

6. Under which of the following headings might you list the animals described in this story?

   a. Tomorrow's Scientists

   b. Endangered Species

   c. Future Forecasters

The Octopus—a Sea Creature

In 1896 a large sea creature washed up on a beach in Florida. Some scientists believed the creature was a giant octopus. It weighed more than 8,250 pounds. And it had arms that were about 82 to 99 feet long. Since then, no one has found an octopus of this size.

There are about 50 different kinds of octopuses that live in the world's oceans. The octopus belongs to a large group of animals called mollusks (mŏl'əsks). *Mollusk* comes from the Latin word meaning "soft."

Mollusks include snails, mussels, and oysters. But, unlike these animals, the octopus does not have a hard protective shell. In fact, the octopus is boneless like a worm and has a soft, bag-like body. Eight long, arm-like parts extend from its body, and the octopus uses these arms to pull food into its mouth. The mouth contains powerful jaws that come to a point like a bird's beak. The octopus hunts and feeds on shrimp, crayfish, crabs, and mussels.

Today, the largest known octopus lives in the northern Pacific Ocean. It weighs up to 120 pounds and has an arm spread of 23 to 33 feet. But most octopuses are smaller, and some are no bigger than a human's fist!

Will a giant octopus like the one found in Florida ever be seen again? No one can say for sure.

1. The word that comes from the Latin word meaning "soft" is _____ .

2. In 1896 a sea creature that weighed more than _____ washed up on a beach in Florida.

    a. 27 to 33 pounds

    b. 55 pounds

    c. 8,250 pounds

3. The octopus hunts and feeds on

    a. snails and slugs.

    b. crayfish and crabs.

    c. salmon and oysters.

4. According to the story, snails, mussels, and oysters have

    a. powerful jaws.

    b. large eyes.

    c. hard shells.

5. The arm-like parts of the octopus are also used for defense.

    a. True

    b. False.

    c. The story does not say.

6. Under which of the following headings would you look to find out more about the octopus?

    a. Boneless Creatures of the Sea

    b. Sea Animals with Protective Coverings

    c. Sea Creatures of Long Ago

No Longer a Threat

The mountain lion is a member of the cat family. It is called by a variety of names, such as *cougar*, *puma*, or *panther*. For many years, mountain lions were regarded as threats to humans and other animal life. The great cat was branded a killer, and hunting it was legal. In fact, a *bounty* (boun'tē), or reward, was given to people for each mountain lion they killed. By the 1950s, the great cat was put on the list of endangered *wildlife* (wīld'līf'). Wildlife refers to those plants and animals living in their natural surroundings.

In the early 1960s, the state of Idaho hired Maurice Hornocker, a scientist. Hornocker studied a group of mountain lions. The lions lived in the Idaho Primitive Area. For ten years, Hornocker captured, studied, and released many mountain lions. He attached a radio collar to the big cats. A beeping signal from the collar helped Hornocker to track the animals. He proved that only the

most bloodthirsty mountain lions attacked sheep, cattle, or deer. And these attacks were rare. He also found that most adult mountain lions do not live or hunt together. Hornocker believed that because the great cats are loners, hunters were not able to kill off the entire species.

Hornocker's work helped change public opinion. States have transferred mountain lions from the *bounty list* to the *game animal list*. Now, the hunting of mountain lions is being controlled.

1. The word used to describe plants and animals that live in their natural surroundings is _____.

2. What other names for a mountain lion are mentioned in the story?

3. According to Hornocker, what saved the big cats from being completely wiped out?

    a. They do not live or hunt together.

    b. They are able to protect their young.

    c. Most states put them on the bounty list.

4. Because it was on the *bounty list,* the mountain lion was

    a. nearly killed off by hunters.

    b. no longer considered a killer.

    c. required to wear a collar.

5. Hornocker studied the big cats mainly because

    a. he wanted to test his radio collar invention.

    b. Officials in Idaho wanted to know more about them.

    c. he wanted to hunt them for the reward.

6. Under which of the following headings would you list the mountain lion?

    a. Animals on the Bounty List

    b. Members of the Panther Family

    c. Animals That Live in Groups

Why Muscles Move

If you had no muscles, nothing in your body would move.

Without muscles (mŭs'əlz), you would not be able to run, eat, or blink your eyes. Your heart would not beat, and the blood would not be carried through your veins. A muscle is a kind of body tissue (tĭsh'oo) made up of fibers, or thin, thread-like strands. There are more than 600 muscles in your body. They all do their work by shortening, or contracting, and relaxing. When muscles contract, they tighten and pull.

This pulling action works because no muscle ever works alone. All muscles work in pairs or groups. Study the picture below. The top muscle in the arm is called a biceps (bī'sĕps'). To raise the upper arm, the biceps contracts and pulls the arm up.

The lower muscle shown is called a triceps (trī'sĕps'). To lower the arm, the triceps contracts and pulls the arm down. When the biceps contracts, the triceps relaxes. When the triceps contracts, the biceps relaxes. These two muscles work together to raise and lower the upper arm.

All muscles have nerves (nûrvz) that carry messages to and from your central nervous system. Muscles like those in your arms and legs work in answer to messages from your brain. These are called voluntary (vŏl'ən tĕr'ē) muscles. Other muscles, like those in your heart and stomach, work without your thinking about them at all. These are involuntary muscles.

1. A kind of body tissue made up of thin, thread-like strands, or fibers, is called a _____.

2. Messages from the brain are carried to the muscles through the _____.

3. Muscles always work in _____.

4. The muscles that help carry the blood through your veins work like the muscles in your

   a. chest and back.

   b. heart and stomach.

   c. arms and legs.

5. Which of the following is a *true* statement?

   a. Your muscles work in answer to the needs of your body.

   b. Voluntary muscles work without your having to think about them at all.

   c. When muscles contract, they get longer.

6. Muscle action is very much like the action of a

   a. seesaw.

   b. rubber band.

   c. bouncing ball.

Studying the Lunch Bunch

That question is on the minds of many food experts and nutritionists. The food we eat should help us grow and develop properly. Nutritionists study food to see if it is healthful, or *nutritious* (noo trĭsh'əs).

What we eat at lunch can satisfy about one-third of our daily food needs. So experts are studying school lunches. They believe that a healthful lunch should have many of the *nutrients* (noo'trē ənts) a student needs. A nutrient is the part of a food used by the body for growth and energy. Protein, fats, carbohydrates, vitamins, minerals, containing calories, or food energy, are examples of nutrients.

Not all foods contain the same amount or kind of nutrients. Calories supply your body with energy. Protein is used by your body to replace and build new cells. Vitamins, like niacin, help you to grow. Vitamin D keeps bones healthy. Minerals, such as iron, move oxygen through the body.

Nutritionists have learned that too much fat, sugar, and

salt are not good for you. School dietitians keep this in mind when they plan the lunch menu. But even a well-balanced lunch will not nourish the student who just does not eat. So nutritionists are looking for ways to make school lunches taste better. Some school dietitians are adding nutrients to favorites such as pizza, tacos, and milk shakes.

1. The part of a food that is used by the body for growth and energy is called a _____.

2. A person who studies food to see if it is healthful is a _____.

3. Food experts have learned that too much _____ and sugar is not healthful.

4. Which is the most important reason for eating different kinds of food?

    a. so you can learn what different foods taste like

    b. because foods contain different nutrients in different amounts

    c. because the more you eat, the better you feel

Use the table below to answer questions 5 and 6.

FOOD GROUP	FOOD	NUTRIENT
I. Bread and Cereal	Whole wheat bun	Carbohydrate
II. Milk products	Milk, Swiss cheese	Calcium, protein, fat
III. Meat, fish, poultry	Hamburger	Protein, fat
IV. Fruits, vegetables	Lettuce, orange	Vitamins A and C

5. The name of the Food Group to which cheese belongs is _____.

6. According to the table, Food Groups II and III both provide these nutrients: _____ and _____.

A Home for the Birds

What good is a dead tree?

Some animals make their homes in trees. But fire can often destroy or thin out forests. And, in some places, air pollution can kill trees and other plant life. As a result, animals have a problem finding food and a place to live. But, for some animals, these problems can be solved by a dead tree.

In the United States, about 85 different kinds of birds build their nests in *snags* (snăgz). A snag is the stump that is left after the branches have been cut or torn off a tree. The North American *chickadee* (chĭk'ə dē) is one bird that often makes its home in a snag. First, the bird looks for a place where the dead wood has become soft and rotten. Then, it cleans out the soft, dead wood from the snag. The bird makes its nest in the hole that remains. Other

animals that do this are raccoons, chipmunks, and flying squirrels.

Snags provide food for many kinds of animals. Woodpeckers, for example, dig for insects that make their homes in snags. The woodpecker is unusual. It is able to dig for insects in both live and dead trees with its hard, sharp beak.

Snags are also found sticking up out of lakes and rivers. They can be a danger to ships and boats. But deciding whether or not a snag should be cut down or burned is not an easy task.

QUESTIONS

1. After the branches of a tree have been cut or torn off, the remaining stump is called a _____.

2. The woodpecker is unusual because it
   a. makes its home in snags.
   b. can dig into both live and dead trees.
   c. does not eat insects the way other birds do.

3. In addition to being a home for many animals, a snag may also provide _____ for the animals.

4. What does a bird do *first* if it wants to live in a snag?
   a. It cleans out the tree's soft, dead wood.
   b. It looks to see if other animals are living in the snag.
   c. It finds a place in the snag where the wood is rotten and soft.

5. What is the *main idea* of this story?
   a. Forests are often thinned out by fire.
   b. Dead trees can be put to good use.
   c. Animals have problems finding food.

6. Of the following groups of people, who do you think would favor burning a snag?
   a. sailors
   b. bird-watchers
   c. forest rangers

An Amazing Little Weed

To some, it is a tasty meal. To gardeners, it is a pest. In spring, it is seen almost everywhere. This bright yellow wild flower is called a dandelion (dăn′dl ī′ən). The dandelion spreads quickly and is hard to control. Its seeds are carried long distances by the wind, and its roots grow to 3 feet. This plant is a perennial (pə rĕn′ē əl). This means that its leaves and flowers die, but its roots live from year to year.

The word *dandelion* means "lion's tooth." The leaves of the dandelion are notched and look very much like lions' teeth. The tender young leaves of the dandelion are often used in salads. They are also cooked and eaten as a vegetable. Dandelion leaves are rich in minerals and vitamins A and B.

Animals and insects eat the roots of the dandelion, which are cooked and eaten by people, too. Because they grow so deep, the roots reach minerals and moisture not reached by other plants. The minerals and moisture travel up the dandelion's hollow stem. When the plant does not use them up, the minerals and moisture go back into the earth. There, they get absorbed by other plants.

When the dandelion's flowers die, a fluffy gray ball is left. This ball holds the plant's seeds. Birds like to eat the seeds. People like to blow on them and make a wish. Did you ever wish on this amazing little weed?

1. When plants have leaves and flowers that die but roots that live on from year to year they are called _____.

2. According to the story, the seeds of the dandelion are eaten by
    a. people.
    b. birds.
    c. insects.

3. Dandelion leaves can be cooked and eaten as a vegetable.
    a. True
    b. False
    c. The story does not say.

4. Why does the dandelion spread so quickly?
    a. because it absorbs all the minerals in the earth
    b. because its seeds are carried by the wind
    c. because it is eaten by birds and insects

5. The dandelion gets its name from the look of its
    a. leaves.
    b. roots.
    c. stems.

6. Which of the following words describes the dandelion?
    a. harmful
    b. useful
    c. uncommon

The Desert Food Chain

The desert is full of all kinds of life.

It is true that a desert has little rainfall, but it is full of life. To *survive* (sər vīv′), or stay alive, all the living things in the desert must depend on one another. These living things form a food chain. The food chain begins with the sun. Animals cannot use the sun's energy directly to make food. But green plants can use the sun's energy to make *glucose* (glōō′kōs′), or plant sugar. The glucose is used to make all the nutrients a plant needs to grow and reproduce.

In the desert and elsewhere, green plants make their own food. They do this by a process called *photosynthesis* (fō′tō sĭn′thĭ sĭs). Photosynthesis works in this way. First, the plants *absorb* (ăb sôrb′), or take in, what little moisture there is in the desert. Then they combine the moisture with *carbon dioxide gas* (kär′bən dī ŏk′sīd′ găs) from the air and light energy from the sun to make glucose.

The smaller animals, such as snakes, lizards, mice, birds, and insects, get their food and water by eating the plants. The next higher level in the desert food chain includes owls and foxes. These larger animals get food and water by eating the smaller animals. Each plant or animal in the desert is an important link in the chain of survival.

1. The process by which plants make their own food is called _____.

2. All living things in the desert must depend on each other in order to _____.

3. The word *photosynthesis* means "putting together with light." This means that plants use light energy from the _____ during photosynthesis.

4. Moisture is not necessary for photosynthesis to take place.

    a. True

    b. False

    c. The story does not say.

5. From the list below, find the missing link in the desert food chain.

    Sun ⟶ _____ ⟶ Mice ⟶ Foxes

    a. Owls

    b. Plants

    c. Birds

6. Suppose you are lost in the desert. You have no water left to drink. Which of the following would be the *best* way to get water?

    a. Take in some carbon dioxide.

    b. Collect some rainwater.

    c. Eat some plants.

Fungi, A Different Kind of Organism

Fungi get their food in different ways.

Organisms such as mushrooms, yeasts, rusts, smuts, molds, mildews, and puffballs are called fungi. Some fungi are *decomposers*. A decomposer gets its food and energy by breaking down the wastes, or remains, of other organisms. Fungi help the environment. They break down organic waste materials that would otherwise build up on the Earth's surface. Other kinds of fungi get their food from living things. These are called parasites.

Fungi live in many different places throughout the world. Most fungi live in places that are moist, dark, and warm. Fungi that live in soil include mushrooms, puffballs, and the water molds. Other fungi such as bracket fungi and various types of mushrooms live on decomposing plant matter such as rotting logs.

Fungi are used to make a variety of products. For example, some yeasts are used in the making of breads. Other fungi are used to make cheeses. Fungi also produce materials that are used for medicines, such as penicillin. This medicine has been used to treat a variety of infections since the 1940s.

However, many fungi are poisonous to humans when eaten. Other fungi cause diseases. For example, Dutch elm disease is caused by fungi that destroyed many of the elm trees in North America and Europe. Fungi also cause damage to corn and potato crops.

1. Another word for *organism* is
    a.   a living thing.
    b.   an animal.
    c.   a plant.

2. According to the story, waste materials are broken down by
    a.   parasites.
    b.   decomposers.
    c.   animals.

3. Which of the following describes fungi?
    a.   harmful and helpful
    b.   only harmful
    c.   only helpful

4. In the story, penicillin is a
    a.   food.
    b.   parasite.
    c.   medicine.

5. If you saw fungi growing on a dead branch of a tree, it would probably be
    a.   bracket fungi.
    b.   water mold.
    c.   yeast.

6. Which of the following is a *false* statement?
    a.   Bracket fungi live in soil.
    b.   Yeast is used to make bread.
    c.   Penicillin is a medicine.

Your Incredible Breathing Machine

Humans can survive only for a few minutes without air.

Air is a mixture of gases. Of all the gases in the air, oxygen is the most important. Without oxygen, the body would not be able to produce the energy it needs to do work.

Air usually enters the body through the nose and passes down into the lungs. After filling the tiny sacs in the lungs, the oxygen in the air passes into the blood. So when you breathe in, or inhale, the air around you, you take oxygen into your lungs. When you breathe out, or exhale, you force carbon dioxide, a waste gas, out of your lungs. This exchange of oxygen and carbon dioxide is called *outer respiration* (rĕs′ pə rā′shən).

There is another kind of respiration, called inner *respiration*. It takes place when the cells in your body take in oxygen from the blood and send carbon dioxide back into the blood.

Breathing is controlled mostly from a nerve center in the brain called the respiratory center. The amount of carbon dioxide in the blood affects this center. Holding your breath, for example, increases the amount of carbon dioxide in your blood. The extra carbon dioxide causes the respiratory center to send out messages, or nerve impulses, to the muscles used for breathing. The nerve impulses make you take a deep breath, even if you don't want to. You breathe more quickly until the amount of carbon dioxide in your blood has returned to normal.

1. The exchange of oxygen and carbon dioxide is called _____.

2. A waste gas that is forced out of your lungs when you exhale is _____.

3. The human body needs oxygen in order to produce _____.

4. Number the following events (1, 2, 3) in the order in which they occur in the story:

    _____ a.   Oxygen fills the tiny sacs in the lungs.

    _____ b.   Air enters the nose.

    _____ c.   Oxygen passes into the blood.

5. What occurs during outer respiration?

    a.   You inhale oxygen and exhale carbon dioxide into your surroundings.

    b.   You exhale oxygen into the tiny sacs in the lungs.

    c.   You inhale carbon dioxide and exhale oxygen into the blood.

Use what you have learned from this story to answer question 6.

6. Your body would most probably react to the lack of air in a small, hot, stuffy room by making you

    a.   cough a lot.

    b.   cry.

    c.   inhale deeply.

EARTH-SPACE SCIENCE

People have long been interested in the beautiful and strange world under the surface of the ocean. Thousands of years ago, Greek and Roman divers explored the Mediterranean Sea looking for shells, pearls, and sponges. Today, modern equipment makes it possible for divers to stay underwater for long periods of time. This diver inhales air from a tank she carries on her back. She exhales air into the water through a tube. Because air is lighter than water, the bubbles of air rise to the surface.

WORDS TO KNOW

Can Earthquakes Be Predicted?

property, an object, such as a building, that is owned by a person or group

beneath, in a lower place

predict, to tell what will happen

Fossils—Clues to the Past

fossils, the hardened remains of plants or animals

ancient, very old

compressed, squeezed down

rare, not often found

ancestors, persons from whom one is descended

Treasures on the Ocean Floor

nodule, a lump of a mineral, or a mixture of minerals

dredged, gathered (up) by net or by suction from the bottom of a river, lake, ocean, etc.

refinery, a factory for refining, or purifying, raw materials

The Milky Way Galaxy

comprehend, to grasp, understand

Finding Extraterrestrial Rocks

extraterrestrial, from outside Earth

meteorite, a piece of rock or metal reaching Earth's surface from outer space

Saturn and Its Mysteries

revolves, goes in circles

investigate, to search into systematically

atmosphere, all of the gases surrounding a planet

microbes, microscopic (only able to be seen with a microscope) organisms

Mass Movement

landslide, the slipping down of a large amount of dirt or rocks on the steep slope of a hill or mountain

gravity, the natural force that makes all bodies or things fall toward the center of the earth

slope, land that slants up or down

bedrock, the solid rock under the surface soil of a hill or mountain

The Sun Makes the Weather

vapor, particles of water floating in the air

Sunspots and Climate

diameter, the straight-line distance from one side of a circle through the center to the other side of the circle

migrate, to move from one place to another

rotate, to turn around

equator, an imaginary circle that divides a sphere into two equal parts

Missing the Mark

spacecraft, a vehicle that travels in space

jets, motors that shoot out strong streams of gases

orbit, a circular course around an object

The Landings on Mars

craters, bowl-shaped depressions

The Most Terrible Storm on Earth

spiral, circling around a central point

eye, the center of a hurricane, usually clear and calm

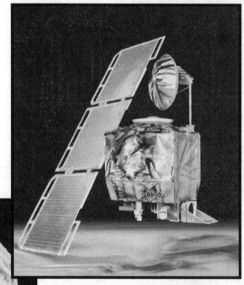

Can Earthquakes Be Predicted?

Will there be a time when earthquakes can be predicted?

Each year there are about 25,000 to 50,000 earthquakes. They take place in many countries. They are caused by movements of rock layers beneath the Earth's surface. Most of the movements are little. They cause no damage. However, some are powerful and damaging.

In 1999 a powerful earthquake occurred in Turkey. More than 16,000 people died. About 38,000 were injured. Thousands were left homeless. In 1995 a powerful earthquake rocked Japan. More than 5,000 people were killed and 26,000 were injured. The property damage was about 100 billion dollars. Earthquakes can destroy lives and leave people homeless. They also damage and destroy buildings, roads, bridges, and water, sewer, and telephone lines.

Many scientists would like to predict when earthquakes will occur. Earthquakes cannot yet be predicted with any accuracy. The next best thing is to develop

earthquake prevention plans to avoid loss of life. One idea is to build stronger buildings and bridges that can withstand earthquakes. Another idea is to make laws to keeping people from living in areas where earthquakes are most likely to happen.

1. Earthquakes are caused by movements of _____ beneath Earth's crust.

2. What does the word *accuracy* mean?

    a. some errors
    b. no errors
    c. many errors

3. According to what you read, most lives lost during an earthquake are due to

    a. highways being destroyed.
    b. buildings fallen down.
    c. fallen trees and telephone poles.

4. To help prevent earthquake damage in a house, which of the following is

    the best idea?

    a. Remove objects from high shelves.
    b. Store objects in boxes.
    a. Unplug all electrical appliances.

Use the table to answer questions 5 and 6.

STRONG EARTHQUAKES		
Year	Location	Approximate Number of Deaths
1920	China	180,000
1939	Turkey	33,000
1970	Peru	66,000
1976	China	240,000
1985	Mexico	9,500
1990	Iran	50,00
1994	California	60

5. Which country had the greatest number of deaths caused by earthquakes?

    a. Peru
    b. China
    c. Iran

6. Which country had the least number of deaths caused by earthquakes?

    a. Turkey
    b. Peru
    c. Iran

Fossils—Clues to the Past

How do scientists learn about prehistoric plants and animals?

Much of what scientists know about the past comes from clues from fossils (fŏs′əlz). Fossils are the remains of ancient plant or animal life. They are often found buried.

One kind of fossil might be the tracks of an ancient animal. The fossil may have formed in this way:

Millions of years ago, an animal left its footprints, or tracks, in damp soil. Soon after, the tracks became buried in mud. They may even have been covered by water. Over time, the earth compressed (kəm prĕst′), or squeezed, the layers of mud into rock. The hardened animal tracks were not seen for millions of years.

Most fossil tracks are not found by accident. They are searched for in areas where there is sandstone or other compressed rocks. Searchers have found all kinds of fossils. These include prints of 300 million-year-old ferns as well as tracks left by worms, birds, and dinosaurs.

In Africa, in 1977, Dr. Mary Leakey made a rare discovery. Under a thin layer of soil, the scientist found the 3.6 million-year-old fossil tracks of two ancestors of humans. The tracks looked as if they had been made yesterday!

1. The word in the story that means the remains of ancient plant or animal life is _____.

2. According to the story, rock was formed when layers of mud were
    a.   buried.
    b.   compressed.
    c.   frozen.

3. Most fossil tracks are discovered by accident.
    a.   True
    b.   False
    c.   The story does not say.

4. The ancient fossil tracks discovered by Dr. Mary Leakey in 1977 were very clear. How is this possible?
    a.   They had been kept under water for many years.
    b.   They had been protected by the thin soil covering.
    c.   They had been dug up very gently.

5. The title leads you to believe that examining fossils is a way of
    a.   bringing prehistoric animals back to life.
    b.   protecting plants from dying.
    c.   learning about life in the past.

6. Based on what you have read, you could say that a scientist might use a fossil the same way _____ might use a _____.
    a.   a carpenter, hammer
    b.   a detective, fingerprint
    c.   an artist, paintbrush

Metal Treasures on the Ocean Floor

What treasures are on the ocean floor?

They can be as big as tennis balls. Some are pea-sized or smaller. What are they? They are deep-sea *nodules* (nŏj′ ōōlz). A nodule is a lump of a mineral or a mixture of minerals. Nodules are found on the ocean floor. They are made mostly of metals like copper, iron, and gold. Most nodules lie at depths of around 4 miles. Scientists think there may be many nodules in the northern Pacific Ocean.

Why would we go to the trouble of mining the ocean? The metals in nodules are growing scarce on land. In the twenty-first century, nodule mining may become an important source for metals. Experiments with nodule mining have already started. Two ways to bring nodules to the surface use a mining ship. One way is to use buckets. The buckets are lowered by heavy wires. The buckets are

then dragged across the ocean bottom. When they are filled, the buckets are brought up. The nodules are then dumped into the ship's hold. A second way nodules are mined is by lowering long pipes from the ship down to the ocean floor. Then the nodules are sucked up through the pipes and stored on board. A concern is that any method that becomes widely used should be one that would not hurt the ocean floor.

When mining ships return to port the nodules are transported to a *refinery* (rĭ fī′nə rē), a type of factory. The metals are removed from the nodules and then sold.

1. The word *nodule* is used to describe a small lump of

    a. gold.

    b. copper.

    c. minerals or mixture of minerals.

2. Nodules are found

    a. many miles down on the ocean bottom.

    b. inside caves by islands under the ocean.

    c. inside pipes near coral reefs.

3. Some scientists think that the _____ Ocean may be a rich source of nodules.

4. Ocean nodule mining is becoming more important because

    a. humans have used up a lot of the metals found in the Earth.

    b. the metals found on the ocean floor are more useful than those found on land.

    c. it is cheaper to mine the ocean than the land.

5. Whichever method is used to mine nodules, the *last* step always involves

    a. sucking up the nodules through pipes.

    b. removing the metals from the nodules.

    c. scraping the nodules off the ocean floor.

6. If you took a job as a nodule miner, why wouldn't you be able to dive down and get the nodules?

    a. The buckets of metal would weigh too much.

    b. Divers can't dive deep enough because the water pressure would crush them.

    c. It is too dark that far down to see the metals.

The Milky Way Galaxy

What is a galaxy?

The ancient Greeks enjoyed looking at the night sky. They found that a part of the sky was foggy-looking. They called this patch of foggy sky galaxias (gə lăk̄sē äs), which means "milky." The Greeks were looking at the group of stars that we now call the Milky Way galaxy.

The Milky Way galaxy stretches across the sky in the shape of a large wheel. It has many billions of stars in it. Our sun is only one little star out near the edge of the galaxy. When the moon is not shining, the galaxy does not look milky. With a pair of binoculars, it is possible to see stars in the Milky Way.

Earth is a part of the Milky Way galaxy. To discover what our own galaxy is really like, scientists study other galaxies through powerful telescopes.

They know from these studies that galaxies turn and that stars circle around inside them. Planets may then circle around the stars, just as Earth revolves around its nearest star, the sun.

It is difficult to comprehend, or understand, how huge the Milky Way galaxy is. It is even more difficult to comprehend all the stars in it. And it is most difficult of all to comprehend that the Milky Way is only one of millions of galaxies in our universe.

1. The Greek word *galaxias* means _____.

2. The Milky Way galaxy is in the shape of a _____.

3. Scientists have learned from their studies that

    a. the stars circle around the planets.

    b. galaxies turn and stars circle around inside them.

    c. galaxies far from the Milky Way have no stars.

4. In the Milky Way galaxy, the sun is the star that

    a. is closest to Earth.

    b. revolves around Earth.

    c. is located in the center of the galaxy.

5. Scientists use telescopes to see other galaxies because the other galaxies are

    a. so crowded.

    b. so dark.

    c. so far away.

6. Which of the following conditions would be *best* for seeing the stars in our galaxy?

    a. a night when the galaxy's milky look is seen clearly

    b. a clear, moonless night

    c. early evening before the sky gets too dark

Finding Extraterrestrial Rocks

In the late 1960s and during the 1970s, U.S. astronauts went to the moon. They brought back rocks and dust for scientists to study. The trips to the moon are examples of sample return missions. People traveled into space and came back with samples to study. Other sample return missions are planned. They will use robots instead of people. The robots' jobs will be to collect rocks and dust from comets, asteroids, and the moons of other planets. Going into space is one way to get rocks and dust from other planets. Did you know you could find these things right here on Earth?

Would you believe a rock from Mars was found in a man's backyard in California? It's true! Pieces of Mars fall to the Earth in the form of meteorites and have been found in many parts of the world. These meteorites were created

long ago. Comets or asteroids hit Mars with such force that they blasted pieces into space. After millions of years, some of these pieces of Mars found their way to the Earth.

In 1996 scientists said they had found what might be evidence, or proof, of Martian life in one of these meteorites. While other scientists do not agree with their ideas, everyone does agree that pieces of Mars can be found—right here on Earth.

1. A *sample return mission* means that

    a.   men or robots go into space and came back with rock samples.

    b.   rocks from other planets can be found right here on Earth.

    c.   scientists have never been able to study rocks from other planets.

2. Pieces of Mars are blasted into space and eventually find their way to Earth in the form of _____.

3. Martian life has been discovered in a meteorite found on Earth.

    a.   All scientists agree that this is true.

    b.   All scientists agree that this is false.

    c.   Some scientists think so, but many do not.

4. According to the story, which statement is *true*?

    a.   Martian rocks took only minutes to reach the Earth.

    b.   Martian rocks took a few months to reach the Earth.

    c.   Martian rocks probably took millions of years to reach the Earth.

5. Answer this question based on the information in the story: Where can you find Mars rocks on Earth?

    a.   Only in people's backyards in California.

    b.   There are no Martian rocks on Earth.

    c.   Mars rocks have been found in many places on Earth.

6. Which do you think might be the *least expensive* way to find rocks from Mars?

    a.   Through a sample return mission that goes to Mars.

    b.   By finding them here on Earth.

    c.   By going into deep space to find them on their way to Earth.

Saturn and Its Mysteries

The ringed planet has long been a mystery to astronomers.

Saturn, the ringed planet, is one of the more interesting bodies in our solar system. Second only to Jupiter in size, Saturn is 9 1/2 times the size of Earth. It is the sixth planet from the sun and is about 10 times farther away from the sun than Earth. Saturn receives a small amount of the sun's heat and light. The earth revolves (rĭ vŏlvz′), or circles, around the sun in one year. But it takes Saturn about 30 Earth years to make the trip.

For several years, astronomers have known about the many rings that revolve around Saturn. In recent years, several spacecraft have flown by Saturn and sent back new data on the strange rings. Saturn has hundreds or possibly thousands of ringlets. Only a small space separates one from another. The ringlets are in nine different groups. Larger spaces separate them. The rings appear to be made up of many layers of snowball-sized pieces of ice or ice-covered rocks.

Saturn has eighteen moons—at least as far as we know now. Spacecraft from Earth, *Pioneer II*, *Voyager I* and *Voyager II*, have given us a good deal of knowledge about the planet. The *Cassini Probe*, a nuclear-powered spacecraft that will reach Saturn in 2004, will hopefully tell us even more. One of its missions will be to investigate Saturn's atmosphere. Another will be to investigate the atmosphere and even the surface of Saturn's largest moon, Titan. It is believed that there may be water on Titan and possibly even living microbes.

1. The word in the story that means "to circle, or, to go around" is

    a. *detect*.

    b. *launch*.

    c. *revolve*.

2. Information from spacecraft shows hundreds or possibly thousands of _____ around Saturn.

3. The rings of Saturn appear to be made up of _____.

4. A year on Saturn is equal to _____ year(s) on Earth.

    a. 1

    b. 4

    c. 30

5. Saturn does not receive as much heat and light from the sun as Earth does, because Saturn

    a. circles the sun more slowly.

    b. is farther away from the sun.

    c. has more moons than Earth.

6. From the story, you can tell that

    a. Janus is the largest of Saturn's moons.

    b. Saturn's rings are mostly cloud-like substances.

    c. It would be difficult to pass through Saturn's rings.

Mass Movement

What causes landslides?

Suddenly, the side of a hill shifts, sending rocks and mud roaring down below. Homes and property are buried or swept away. What is this frightening event, and what caused it? The event is a landslide, one example of what scientists call Mass Movement.

In Mass Movement, large amounts of Earth's surface slide down the slope of a hill or mountain. Sometimes mud and rocks move very slowly.

In a landslide, the materials are bedrock or loose rock. They move suddenly, rushing down the slope of the hill or mountain. The main cause of these rock falls is gravity acting on a very steep slope. But there are other things that also cause landslides.

Rivers or ocean waves may make the slope steeper. Heavy rains or the melting of heavy snow may weaken the soil or rocks. The stress of an earthquake may also make weak slopes fall. Traffic and the shaking of heavy road-building machinery may also weaken the slope.

Landslides happen in every state in the United States. Any area that has large amounts of very weak or broken rocks that rest on a steep slope can have land-slides. But they happen most often in mountain areas.

Scientists keep records and maps that show the areas where landslides are most likely to happen. The physical causes of many landslides cannot be removed. But control of land use and good engineering practices can reduce the hazards of landslides.

1. A landslide is one example of _____ _____.

2. The main cause of landslides is _____ acting on a very steep slope.

3. Why might rivers or ocean waves make a slope steeper? _____
   _____

4. Scientists keep records and maps that show

   a. what causes landslides.

   b. where landslides are most likely to happen.

   c. how gravity can cause landslides.

5. Which of the following is a physical cause that cannot be controlled?

   a. heavy traffic on mountain roads

   b. farming on the side of a hill or mountain

   c. the stress of an earthquake

6. Why is it important to know where landslides are most likely to happen?

   _____

   _____

   _____

The Sun Makes the Weather

The next time you are caught in a sudden rainstorm without your umbrella, blame the sun. That's right. The sun not only makes sunny days, it also makes lots of rainy ones. Windy days and snowy days can also be gifts from the sun. This can be explained by the water cycle (wô′tər sī′kəl). The water cycle is the movement of water from the ocean into the air and back to the ocean again.

The water cycle starts when the sun shines down directly on the ocean. This heat causes some of the ocean water to change into water vapor (vā′pər). Fog and steam are kinds of water vapor. Some of the sun's energy gets trapped in the water vapor. This causes the heated vapor to rise up above the ocean. Sometimes clouds are formed from the vapor. The vapor can also become a fast, moist wind that moves across the Earth. As the wind moves, it may drop its moisture as rain, sleet, or snow. The water cycle is complete when water in the form of rain or snow falls back to the Earth. Some of this water returns to the ocean.

The winds get most of their energy from the sun. Sometimes the sun will give off more energy than usual. As this energy nears Earth, it can throw the winds off course.

Then unusual weather may occur. Rain heading for Maryland, for example, may fall in New York.

1. The movement of water from the ocean into the air and back to the ocean is called the _____.

2. What happens to water vapor when some of the sun's energy gets trapped in it?

    _____

    _____

3. A cloud is one form of water vapor.

    a. True

    b. False

    c. The story does not say.

4. Earth's main weather maker is the

    a. wind.

    b. rain.

    c. sun.

5. The wind loses most of its water vapor when the moisture

    a. falls as rain or snow.

    b. dries up in the air.

    c. moves higher into space.

6. Based on what you have read, which of the following *might* happen because of a sudden blast of energy from the sun?

    a. a moist breeze in Florida

    b. a snowfall in Florida

    c. a sunny day in Florida

Sunspots and Climate

Is sunspot activity related to climate changes on Earth?

Scientists called astronomers study the sun. One of their studies includes observing sunspots on the sun's surface. Sunspots appear as dark spots on the surface of the sun. The dark spots, some as large as 50,000 miles in diameter, move across the surface of the sun.

The greatest number of sunspots occurs about every 11 years. The sunspots migrate from the sun's poles to its equator. Some last for several days. Large ones may last for several weeks or months. The movement of the sunspots shows that the sun rotates faster at its equator than at its poles. Sunspots at the equator take about 27 days to go around the sun, as seen from Earth.

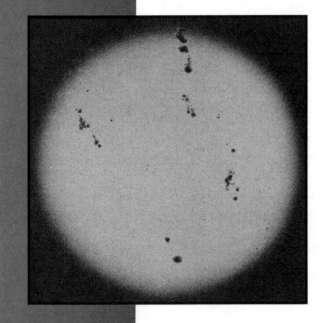

Is Earth's climate affected by sunspot activity? Maybe. Between 1640 and 1710, a low number of sunspots were recorded during that time. The climate in North America was very dry, and many places in Europe had very cold temperatures.

However, more studies need to be done to find out if Earth's climate is connected to sunspot activity. If it is, maybe scientists will be able to predict when the next Ice Age will begin.

1. The dark areas on the sun's surface are known as _____.

2. The greatest number of sunspots occurs

    a. every 11 years.

    b. every 22 years.

    c. every month.

3. From 1640 to 1710 there were a low number of sunspots and

    a. very warm temperatures on Earth.

    b. very mild temperatures on Earth.

    c. very cold temperatures on Earth.

4. The movement of sunspots shows that the sun rotates

    a. faster at its equator than at its poles.

    a. slower at its equator than at its poles.

    c. the same at its equator and at its poles.

Use the table to answer questions 5 and 6.

Sunspot Activity

YEAR	1986	1987	1988	1989	1990	1991	1992	1993	1994
NUMBER OF SUNSPOTS	14	29	100	159	147	145	94	54	31

5. What year did the most sunspots occur?

    a. in 1988

    b. in 1989

    c. in 1990

6. When did the lowest number of sunspots occur?

    a. in 1994

    b. in 1986

    c. in 1987

Missing the Mark

Do you know how to change pounds of force into metric newtons of force? Just multiply the pounds by 4.45 and you'll have newtons. It's an important thing to know. It's key when you are setting the speed of a rocket jet. It's something that NASA forgot when they were setting the steering directions for the Mars Climate Orbiter spacecraft.

Spacecraft are steered right, left, up, and down by *thrusters*. Thrusters are small jets attached to the sides of the craft that produce short bursts of power. When the thruster jet fires in one direction, the craft moves in the opposite direction. The more powerful the thruster's jet burst, the faster and farther the spacecraft moves.

Those who built the Mars Climate Orbiter spacecraft had marked the thruster controls in pounds of force.

Using those settings, the NASA engineers set the thrusters for what they thought were newtons, but were actually pounds. Thus the thrusters were actually set to fire bursts more than four times as powerful as they should have been. As a result, the Mars craft ended up in an orbit 56 miles too close to Mars. It disappeared into the Mars atmosphere and was never heard from again.

1. A pound of force is

    a.   one newton.

    b.   equal to a newton.

    c.   more than four newtons.

2. Thrusters

    a.   steer the spacecraft.

    b.   power the spacecraft.

    c.   destroy the spacecraft.

3. A spacecraft moves

    a.   in the same direction as the thruster jet.

    b.   in the opposite direction from the thruster jet.

4. The greater the force of the steering thruster jet,

    a.   the farther the craft will move in one direction.

    b.   the faster the craft will go.

    c.   the faster the craft will stop.

5. The thruster settings caused the Mars spacecraft to

    a.   stop.

    b.   crash on Mars.

    c.   get too close to Mars.

6. If the Mars craft thruster settings had been marked in newtons and NASA thought the settings were in pounds, what would have happened?

    a.   Nothing would have happenned.

    b.   The craft would have orbited too high over Mars.

    c.   The craft would have crashed.

The Landings on Mars

The 1976 *Viking* landings on Mars gave us a close look at the planet's surface. Dust covers the surface of Mars. And the planet has great, swirling dust storms. The photographs sent back by the Viking spacecraft also showed that the surface of Mars is not smooth. It is covered with rocks, craters, deep canyons, and high mountains.

Mars has a reddish color when seen through a telescope. For this reason, astronomers have called Mars the red planet. Astronomers are scientists who study the planets, stars, and other heavenly bodies. The temperature on Mars is always very cold, dropping to about -200°C at night near the poles. Even during the Martian summer, the temperature probably does not get much above 0°C. On the Celsius scale, zero degrees is the freezing point for water.

No water has been seen on Mars. But astronomers

believe that there may be water below the planet's surface. The Mars Global Surveyor (MGS) landed on Mars in 1997. It may help astronomers find out. The mission of the MGS was to map the entire surface of the planet and the area beneath it. The photographs sent back from MGS were far more detailed than the earlier *Viking* photographs. MGS pictures could show water beneath the surface of Mars. And where there is water, there could be life!

1. Scientists who study the planets, stars, and other heavenly bodies are called _____.

2. The temperature on Mars probably never gets much higher than _____.

3. The surface of Mars is covered with _____, _____, deep canyons, and high mountains.

4. What makes some scientists think that there might be life on Mars?

    a. The planet's surface is not smooth.

    b. The possibility that water might be under the planet's surface.

    c. The great amount of water found at the planet's poles.

5. The mission of the Mars Global Surveyor is important to scientists because it may

    a. show water beneath the surface of the planet.

    b. explain why the surface of Mars is smooth.

    c. prove why Mars is so cold.

6. If you placed a pan of water on Mars, the water would probably

    a. boil.

    b. freeze.

    c. evaporate.

The Most Terrible Storm on Earth

A hurricane is a windstorm with enough force to blow down buildings, uproot trees, and tear ships from their anchors.

This kind of storm begins over warm ocean waters. It can be 288 miles wide, with winds that spiral upward like smoke from a chimney. The eye (ī), or center, of the storm is very calm. But the winds that spiral around the eye can reach speeds of up to 144 miles per hour. These winds carry thick clouds and heavy rains.

In the Northern Hemisphere, the winds of a hurricane spin counterclockwise (koun'tərklŏk'wīz'). Counterclockwise means "in the opposite direction from the hands of a clock." In the Southern Hemisphere, the winds spin clockwise. Many of the hurricanes that reach the Eastern Coast of the United States often start in the southern part of the Atlantic Ocean. They move over water at about 14 miles per hour. Hurricanes travel west, then north, then east. It is this last turn to the east that often brings them to the coast.

Nearing land, the winds grow stronger. They push up huge mounds of water, called a storm surge (sûrj). The stronger the winds, the higher the surge. The surge is the most dangerous part of the storm. It brings high tides and sweeping, pounding waves. Altogether, the "killer" winds, flood rains, and surge of the hurricane make it the most terrible storm on Earth.

1. The center of a hurricane is called the _____.

2. The most dangerous part of a hurricane is the _____.

3. All hurricanes start over _____ waters.

4. The winds of a hurricane may look like

    a. the hands of a clock.

    b. smoke from a chimney.

    c. the roots of a tree.

5. Which of the following choices shows the direction in which a hurricane moves?

    a. East ⟶ North ⟶ West

    b. West ⟶ North ⟶ East

    c. North ⟶ East ⟶ South

6. If a hurricane were heading north at 14 miles per hour (mph) and you were 96 miles to its east, you might still be directly in its path, because

    a. the storm can have winds of up to 288 miles per hour.

    b. the storm can be 144 miles wide.

    c. the storm has winds that spin clockwise.

PHYSICAL SCIENCE

There are clues in this picture that tell you that this building was first hot and then cold. What do you think happened to the building? Why was there so much water on the building? What took the heat out of the water and changed it to solid ice? Where will the heat come from to melt the ice?

WORDS TO KNOW

Up, Up, and Away: Ballooning!

launch, to send up

helium, a gas that is lighter than air

gondola, a lightweight basket hooked to the bottom of a balloon

A Long Flight for People Power

airfoils, aircraft parts or surfaces that control direction, lift, or propulsion of the craft

One Boy's Dream

fascinated, strongly interested

electrons, tiny charged parts of atoms

transmitting, sending from one place to another

cathode ray tube, a vacuum (no air inside) tube that shoots a beam of electrons at a screen

Sonar—Finding Your Way with Sound

sonar, from sound navigation (and) ranging. A device for detecting objects underwater by reflection of sound waves

echo, repeated sound by sound waves bouncing back from objects

Energy Sources

source, a place or thing from which something comes; point of origin

fuel, anything consumed to produce energy

imported, brought in from outside a country

nuclear power, energy released by the splitting of atoms

geothermal, the interior heat of the Earth

Old Fashioned Power at Work

centuries, time measured in hundred-year time frames

labor, work, physical exertion

horsepower, the power exerted by a horse in pulling

Fiber Optics

original, in the beginning, first

surgery, medical treatment to repair an injury, deformity, or disease

Sunlight to the Rescue

solar, having to do with the sun

prized, highly valued

dependable, able to be counted on

Dry Up, Cool Down

evaporating, water becoming a vapor

vapor, the gaseous state of any substance, such as water or gasoline, that is liquid under ordinary conditions

Destination Venus

amateurs, persons who do something as a pastime, rather than as a professional

professional, doing a certain kind of work as a full-time job

wildlife survey, a count of specific animals, plants, and sometimes insect in a defined area

Up, Up and Away: Ballooning!

How are people able to travel in a balloon?

Imagine traveling around the world in a balloon! One group of balloonists has almost actually made it.Ballooning is now a popular sport for all ages. To become a balloonist you must learn how to *launch*, or send up, a balloon. Then you must know how to steer it.

There are two kinds of balloons. One type is filled with air. The other is filled with *helium*, a gas that is lighter than air. The first kind rises because the air inside is heated. Hot air is lighter than cold air so the balloon goes up. Helium-filled balloons rise because helium is lighter than air.

For both types of balloons, the work begins on the ground. First, the giant silk or nylon cloth is unfolded and laid out. The opening in the hot air balloon is turned toward the wind. A fan is used to blow air into the balloon. As the balloon fills up, a gas burner is turned on. The burner heats the air inside the balloon. The helium balloon is filled in a different way. A special pipe is connected to the balloon. Then helium gas is pumped into the balloon until it is filled.

A lightweight basket, called a *gondola*, is hooked to the bottom of the balloon. Ropes hold down the balloon so passengers can board the gondola. The gondola also carries little bags of sand as weights. If a balloon starts descending too rapidly, the extra weight is thrown out.

1. The lighter-than-air gas used in balloons is called

    a. methane.

    b. oxygen.

    c. helium.

2. Balloons are usually made of _____ or nylon cloth.

3. Hot-air balloons rise because the air inside is lighter than the air outside the balloon.

    a. true

    b. false

    c. The story doesn't say.

4. What is the purpose of turning the opening of a hot-air balloon toward the wind?

    a. to fill the balloon up with air

    b. to dry off the wet opening to the balloon

    c. to allow the wind to push the balloon up

5. What could be done to make the helium balloon come down *slowly* for a landing?

    a. Unhook the gondola.

    b. Throw out some sandbags.

    c. Let some helium out of the balloon.

6. Suppose you were in a hot-air balloon when the burner broke. What would happen to your balloon?

    a. It would begin to fall.

    b. It would drift off into space.

    c. It would burst into flames.

A Long Flight for "People Power"

A human-powered flight across the English Channel makes history.

Early one morning on a beach in England, a pilot named Bryan Allen would try to fly his plane across the English Channel to France. Twenty-one miles was not a great distance for most planes, but the *Gossamer Albatross* was a human-powered plane. A human-powered plane had never before made such a long flight.

Allen climbed into the plane's cockpit and began to pedal. The pedals were fastened by a chain to a propeller. As Allen pedaled, the propeller blades turned. Then air began to rush across the upper surface of the plane's wings. A strong force built up beneath the wings, pushing the plane up into the air. The propeller blade and the wings of a plane are known as *airfoils* (âr'foilz'). Among other things, airfoils work with the flow of air to create a force strong enough to lift the plane. As Allen continued to pedal, the *Albatross* met strong winds. Finally, the *Albatross*

landed on the French coast nearly three hours after takeoff. Bryan Allen had made the longest human-powered flight in history.

But Allen had help from Paul MacCready, the plane's builder. The *Albatross* was made of hollow plastic rods covered with thin, strong plastic. It weighed only 70 pounds, half of Allen's weight. The power of Allen's legs was just enough to fly the *Albatross*.

1. The propeller blade or wing of a plane is called an _____.

2. The *Gossamer Albatross* flew across _____.

3. What part of the plane caused the propeller to turn?

    a. the pedals

    b. the wings

    c. the rods

4. The trip made by the *Gossamer Albatross* was important because it was the longest trip made by a

    a. plane built of strong plastic.

    b. propeller-driven, lightweight plane.

    c. plane powered by human energy.

5. Which of the following things happened *first*?

    a. Strong pressure built up under the plane's wings.

    b. The propeller began to turn.

    c. Air rushed across the upper surface of the wings.

6. Under which chapter heading might this story be found?

    a. Transatlantic Flights

    b. Record-Breaking Flights

    c. Dangerous Flights in Space

One Boy's Dream: Television!

Where did television come from?

Everybody knows that Thomas Edison invented the light bulb and Alexander Graham Bell invented the telephone. But do you know who invented the television? Most people do not.

In the early 1900s a teenage boy moved to a farm in Rigby, Idaho. He found a stack of science magazines in the attic. He read them and was fascinated with scientific ideas about tiny charged particles called *electrons*. Electrons are part of atoms. He began to imagine a stream of electrons making a pattern on a tube-like rows of crops in the fields. He dreamed of capturing light in an empty jar and then transmitting it line by line on a beam of electrons. He thought this pattern could be changed to make a picture.

It seems unlikely that a 13-year-old boy with little education, no money, or equipment could do what the greatest electric companies in the world could not do. But Philo T. Farnsworth did! He took his ideas and created a cathode-ray tube. He used the tube to shoot electrons at a screen and was able to make pictures. He had combined the power of radio and the movies. Television was born!

It was many years before Philo got his idea to work. The first televisions were very heavy and had very small screens. There were no big screen TVs. All the pictures were in black and white. And there were only a couple of stations at first. It is amazing that all the television shows and videos we have today exist because of a teenager's imagination!

1. Electrons are

    a. tiny parts of atoms.

    b. workers at electric companies.

    c. the electric outlets we plug TV's into.

2. Which is true?

    a. Philo T. Farnsworth invented the telephone.

    b. Philo T. Farnsworth was 13 when he got the idea for television.

    c. Philo T. Farnsworth worked for a large electric company so he had a lot of money and equipment to invent the television.

3. Philo T. Farnsworth imagined

    a. a beam of electrons running from the ground up into the clouds.

    b. creating a television using lightning during a storm.

    c. capturing light in a jar and organizing it like rows of crops in a field.

4. To invent television Philo T. Farnsworth most needed

    a. imagination.

    b. money.

    c. a computer.

5. If you were to invent something that would change the world, like Philo did, you would first need

    a. a college degree.

    b. lots of money.

    c. a dream.

Sonar—Finding Your Way with Sound

Dolphins and humans can find objects underwater by using sound vibrations.

What do dolphins and humans in submarines and ships have in common? They can both find things underwater by using *sonar*. Sonar is sometimes called echo location. How does sonar work? Dolphins can make a wide variety of high-pitched clicking sounds that travel for long distances underwater. When these clicking sounds hit an object, some of the sound will echo back to the dolphin that made them. By listening to the time it took for the sound to echo back, the dolphin can tell how close the object is. It can also tell whether the object is moving and whether or not it is something good to eat, like a fish.

Humans can't make the same high-pitched clicking sounds with their throats. Instead, they rely on machines to send out sounds, and underwater microphones to hear the echoes that bounce back. Although the ways humans and dolphins send and receive underwater signals are different, the results are the same.

There are many uses for sonar. The military uses sonar to find submarines and also to find undersea mines that could damage ships. Scientists are using sonar to create accurate maps of the ocean floor. They also use sonar to locate sunken ships and other interesting areas for undersea exploration. Fishermen and dolphins both use sonar to find schools of fish.

1. What do dolphins and humans in ships have in common?

    a.   They can both see objects at very far distances using their eyes.

    b.   They can both make high-pitched clicking noises with their throats.

    c.   They can both find objects underwater using sonar.

2. When a dolphin listens to the time it takes for a sound to bounce back from an object, it can tell

    a.   how far away the object is.

    b.   what color the object is.

    c.   whether the object is friendly or not.

3. Because humans are different from dolphins, they have to use _____ to send out sounds and _____ _____ to hear the echoes bouncing back.

4. The high-pitched clicking sounds that dolphins make

    a.   are very hard to hear underwater.

    b.   can travel for long distances underwater.

    c.   cannot be heard at all underwater.

5. According to the story, sonar has many uses, including

    a.   helping submarines to hide from enemy ships.

    b.   mapping the ocean floor and finding schools of fish.

    c.   carving undersea mine shafts using sound vibrations.

6. If you were swimming underwater and suddenly heard a series of high-pitched clicks, you might guess that

    a.   there were humans singing underwater.

    b.   there were definitely sharks nearby.

    c.   there were probably dolphins in the area.

Energy Sources

Many different sources of energy are needed to supply electricity.

Every year, we use more electricity. More than one *source* (sôrs) of energy is used to produce, or make, electricity. A source is a place or thing from which something comes. One of the most common sources of energy is coal. About 57 percent of the electricity used in the United States is generated by coal. Coal is a fuel (fyōō'əl). Fuels are substances that are burned to produce energy. About one-third of the world's coal is found in the United States.

More than 45 percent of the oil used in the United States is imported, or brought in, from other countries. Oil is used to produce about 2.5 percent of the electricity used in the United States.

Another source of energy is natural gas. It is used to produce about 9 percent of the electricity used in the United States. Some of this fuel must be imported, too.

There are other energy sources available. For instance, the United States has its own supply of nuclear power and water power. Nuclear energy provides about 20 percent of the electricity used in the United States. Water power provides about 10 percent of United States electricity.

Three other sources of energy are being developed. *Geothermal* (jē'ō thûr'məl) energy is a kind of steam energy that comes from within the earth. Oil shale is another usable source of energy. But solar energy, or sun power, is considered by some to be the most promising energy source.

QUESTIONS

1. In this story, a substance that is burned to produce energy is called a _____.

2. One of the most common sources of energy used to produce electricity is_____.

3. The United States produces about _____of the oil it uses.
   a. 55 percent
   b. 40 percent
   c. 12 percent

4. According to the story, what is one advantage of using water power instead of oil to generate electricity in the United States?
   a. The United States does not have to import water.
   b. Water power is a more common source of energy than oil.
   c. Water power produces better electricity than oil.

Use the table below to answer questions 5 and 6.

Percentage of Electricity Produced by Source in the United States

| | ENERGY SOURCE | | | | |
Year	Coal	Oil	Gas	Nuclear	Water
1980	50.8	10.8	15	11	12
1990	55.6	4.2	9.4	20.5	10
1997	57.3	2.5	9	20	11

5. Between the years 1980 and 1997, the use of gas as an energy source
   a. decreased.
   b. increased.
   c. remained the same.

6. From 1980 to 1997, the *biggest increase* was in the use of _____ as an energy source.
   a. coal
   b. oil
   c. nuclear power

Old-Fashioned Power at Work

More and more people are using muscle to do work.

Muscle power (mŭs'əl pou'ər) is the kind of energy produced by humans and animals. Centuries ago, human muscles did most of the labor (lā'bər), or work, that had to get done. Then, slowly, animal power began replacing human labor. Dogs, oxen, donkeys, water buffalo, elephants, camels, and eventually horses made up the work force. Even today, the power of a car's engine is still measured in terms of its horsepower. In most countries, machines have replaced muscle power and are used to do work. But shortages of gasoline and other energy-producing fuels have made muscle power important again.

Horses and humans are not machines. But they can sometimes be used to do work in place of machines. People use leg and back muscles to bicycle to and from work. More people are using their muscles to do everyday tasks. These include opening cans, mowing lawns, and shoveling snow. Horses are at work at a variety of jobs,

too. In many cities and on several college campuses, police patrol on horseback. Elsewhere, horsepower is used to deliver newspapers and carry packages. Some farmers use horses to haul wood or to cart maple sap.

It is unlikely that most countries will ever return to the old days. But using human and animal labor on the simple jobs can help save expensive fuels for the really big jobs. Let's hear it for muscle power!

1. The term *muscle power* is used to describe the kind of energy produced by humans and _____.

2. In this story, the word *labor* means about the same as the word _____.

3. The story says that horses are used in cities to

    a.   deliver mail.

    b.   haul wood.

    c.   patrol.

4. What do you think made muscle power seem old-fashioned?

    a.   the invention of simple machines

    b.   the shortage of other kinds of energy

    c.   the expense of using animals for work

5. After reading the story, you can conclude that muscle power is

    a.   slowly replacing horsepower.

    b.   getting popular once again.

    c.   not practical for city use.

6. Under which of the following headings would muscle power best belong?

    a.   A New Form of Energy

    b.   A Powerful Fuel

    c.   An Alternate Source of Energy

Fiber Optics

*Super-
highways
of light*

One of the inventions used every day by most people is the telephone. We use it to talk to other people. We use it to send and receive messages over our computers.

But how are signals carried from one phone to another? The original way was by turning the sound of the speaker's voice into a radio signal. This signal went through copper wires to central stations. Then, it went out over more copper wires until it reached the other telephone. This worked very well until people all over the world got phones. Then they started using more telephone lines to get onto the Internet.

Copper wire just couldn't carry enough of those telephone signals. But light travels faster than sound. Scientists began designing a new system. Now telephone systems around the world are changing from copper wire to fiber optic cables. Fiber optic cables are made up of

very thin threads of glass, as thin as a human hair. These threads are called fiberglass. These cables carry laser light waves. Laser light doesn't scatter the way sunlight does. It remains sharp and clear. Fiber optic cables can carry many more telephone messages than copper wires ever could.

Because fiber optic cables are lightweight, they are also being used in modern planes and cars. They are being used in computer laser printers and even in some forms of surgery.

QUESTIONS

1. Very thin threads of glass are called _____.

   a. laser light waves

   b. fiberglass

   c. fiber optics

2. Cables made of fiberglass carry

   a. copper wires.

   b. radio signals.

   c. laser light waves.

3. According to the story, laser light waves

   a. travel faster than sound.

   b. scatter like sunlight.

   c. only work on the Internet.

4. Telephone systems are changing from copper wires to fiber optic cables because

   a. there are not enough phones in the world.

   b. copper wires cannot carry enough telephone signals.

   c. fiber optic cables are easy to make.

5. Fiber optic cables are being used in more and more modern machines because

   a. they are lightweight.

   b. they are inexpensive.

   c. they never wear out.

6. According to the story, it seems likely that someday soon

   a. there will be no more copper telephone wires.

   b. there will be more fiber optic cables than copper telephone wires.

   c. cell phones will replace phones that use cables.

Sunlight to the Rescue

As you may know, deserts are hot and dry and receive a large share of the sun's energy. Parts of the American Southwest are using the energy from the sun, or *solar* (sō'lər) energy, to meet energy needs. While not much heating is needed there, low-cost air conditioning is as prized as a good rain. And energy from the sun can be used to cool as well as heat.

Solar air conditioning has already been put to use. In fact, in Tucson, Arizona, solar homes are being built that will receive up to 100 percent of their heating energy and most of their cooling energy from the sun. As long as most days have some sunshine, there is energy that can be used for heat or air conditioning. But, in most places, after a few sunless days, stored solar energy gets used up. Until the weather clears, a back-up system must be used. The back-up system would run on other energy sources, such as gas or oil.

In deserts, rainy or cloudy days are rare. So areas such as Saudi Arabia, Central Australia, and the American Southwest would need no back-up source of energy. People living in these areas could depend on receiving a high percentage of the possible sunshine. They could store up solar energy for rainy or cloudy days.

Unlike oil, coal, and gas supplies, the supply of sunlight is almost impossible to use up. Although solar equipment is expensive, sunshine is free and comes without transportation costs. In the long run, solar energy is inexpensive and dependable—especially in the desert.

1. The term *solar energy* is used to describe energy that comes from the _____.

2. The story says that solar energy can be used to cool as well as heat _____.

3. Saudi Arabia and the American Southwest are both _____ regions.

4. The main *idea* of this story is that solar energy is
   a. difficult to store.
   b. a good back-up system.
   c. especially practical in deserts.

Use the table below to answer questions 5 and 6.

CLIMATE INFORMATION ON SIX UNITED STATES CITIES-YEAR 1977

City	Total Days with Precipitation (Rain, Sleet and Snow)	Total Number of Clear Days
Burlington, VT	188	33
Chicago. IL	155	91
Dallas. TX	70	135
Fairbanks, AK	138	53
Los Angeles, CA	32	143
Phoenix, AZ	28	210

5. For which U.S. city would solar energy be the *most* practical?
   a. Dallas
   b. Chicago
   c. Phoenix

6. Which U.S. city in the table would most likely need a back-up system as an alternate to solar energy?
   a. Burlington
   b. Los Angeles
   c. Fairbanks

Dry Up, Cool Down

What makes it possible to cool down in the hot sun?

After a swim on a hot day, you come out of the water. You stand in the sun and shiver. Why? The water drops left on your skin are *evaporating*. Heat from the sun turns water into *vapor*. (See page 56.) As the vapor leaves your skin's surface, you begin to cool off. A breeze speeds up evaporation. When the water drops are gone, you no longer get cooler.

People have used this way of cooling themselves for centuries. In towns around the Mediterranean Sea, water was stored in clay pots. The pots were *porous*, not water tight. Water slowly leaked out of the sides. The water on the outside heated up and evaporated. The water on the inside became cooler.

In India, where the dry season is long and very hot, evaporation helped cool homes. Mats made of grass were hung in the doors and windows facing the wind. In the evening the mats were soaked with water. As the heat and wind evaporated the water, cooling began. All night the mats were kept wet. The temperature could drop as much as 30 degrees (F).

Gardeners use cooling by evaporation another way. In a courtyard garden, they place a small fountain or waterfall. Even on hot days the area is cool. A small tree or large plants make shade. But evaporation and water given off by plants make the difference. Ask the cool toad hidden there.

1. The process of changing water into water vapor is called

    a. condensation.

    b. freezing.

    c. evaporation.

2. A pot made of _____ is porous.

    a. glass

    b. clay

    c. steel

3. Evaporation takes place from the _____ of things.

    a. bottom

    b. inside

    c. surface

4. Why do farmers fighting hot and dry weather fear the wind?

    a. Wind holds in the heat.

    b. Wind speeds up evaporation.

    c. Wind blows down trees.

5. On a hot summer day, which would be the coolest place to meet a friend?

    a. in a small garden

    b. near a large fountain

    c. under a large tree

Destination: Venus

It was a hot August day in 1928. Three friends were ready to fly a homemade rocket to the planet Venus. These men were not trained to be scientists. They were only amateurs, but they were very interested in science. They had spent months working on their experiment. The rocket's engine started up. Black clouds of smoke and red flames poured out of the rocket tubes, but the rocket would not lift off. The experiment was a failure, but at least they had tried.

Today amateurs are making important contributions in many areas of scientific study. Many amateurs are used to count plants and animals as part of wildlife surveys. In the field of astronomy, amateurs help by identifying comets and other events in space that they can see using their own telescopes. Professional scientists often use information collected by amateurs as part of their research. In 1996, David Levy, an amateur astronomer, became famous as

the co-discoverer of the comet Shoemaker-Levy 9. While working on a degree in English literature, he decided to study astronomy on his own. He taught himself the math, physics, and research methods he needed to know how to collect accurate information with his telescope.

Amateur science has changed a lot since those three friends built their rocket in 1928. But their determination and interest in science live on.

QUESTIONS

1. What do we call a person who works at something and is very interested in it, but who has no special training?

2. Were the three amateur rocket scientists successful in reaching Venus?

3. In wildlife surveys, amateur scientists contribute by helping to count _____ and _____.

4. In the field of astronomy, amateur scientists contribute by
    a. helping to observe comets and other events in space.
    b. collecting plants and animals for study.
    c. building rockets for the United States Navy.

5. David Levy is famous for
    a. getting his English literature degree.
    b. his help with wildlife surveys.
    c. co-discovering comet Shoemaker-Levy 9.

6. The story suggests that information collected by amateur scientists is
    a. not important, even if it is correct.
    b. more important than anything professional scientists have found.
    c. sometimes important, and often used by professional scientists.

ENVIRONMENTAL SCIENCE

A butterfly stops at a flower to suck up nectar. Most adult butterflies feed only on nectar. They do not harm the flowers. In fact, they help pollinate flowers. Many flowers must have pollen from other blossoms of the same kind of flower to produce fruit and seeds. Along with bees, butterflies are necessary for the flowering of plants and the blossoming and fruiting of trees.

WORDS TO KNOW

Global Warming

global, of the whole world, worldwide

greenhouse, a building with glass sides and roof, for raising plants

greenhouse effect, the warming effect produced by naturally occurring and man-made gasses in the Earth's atmosphere

climate, the regular weather conditions of an area

industrialized nation, a nation whose manufacturing and production of goods is machine or technology based

crop yields, the amount of food an area can produce

predict, to forecast, guess, or prophesy

Drinking Seawater

desalination, the process by which salt is removed from seawater

process, a series of actions, motions, or operations leading to some result

condensed, reduced to a more compact form

The Great Sperm Whale

mammal, warm-blooded animal

prohibit, to forbid, prevent

extinct, no longer exists

Hydrogen, Fuel of the Future?

communities, towns, cities, regions where people live

plentiful, in great quantity, ample supply

Putting the Sun to Work

solar collectors, devices for trapping and using the sun's energy

absorb, to take in, soak in or up

insulated, built to keep heat in

Jane Goodall and Her Work with Chimpanzees

habits, the usual way something behaves, acts

Plants Give a Warning

constantly, always

detect, find

clue, anything that guides or directs in the solving of a problem or mystery

Watching Out for Salamanders

amphibian, a class of animals, including frogs and toads, that live in water and on land

pesticides, poisons used to kill pests, such as insects

lizards, a group of reptiles that include iguanas, geckos, and chameleons

The Arctic Is Warming Up

region, a part of the Earth's surface

drifting, to be carried along on the ocean currents

several, more than one

stretches, large areas

Volcanoes, A Natural Disaster

disaster, a happening causing
    widespread destruction

billow, to send up a great surge
    of smoke or steam

systems, plans, ways of doing
    something

instruments, tools for measuring

satellites, objects revolving around
    another object

Global Warming

Countries around the world are responding to the risks of climate change.

Have you ever seen a greenhouse? This is a building with a glass roof and sides. It is used to raise plants that need a warmer temperature than can be found outside. The glass roof and sides allow sunlight to enter. They also keep heat from escaping.   In many ways the Earth is a natural greenhouse. Water vapor, carbon dioxide, and other gases help to keep in heat, much like the glass panels of a greenhouse. Without this natural warming process, the Earth would be too cold for us to survive.

Before the wide use of things like coal and oil, climate changes occurred naturally. But most scientists now agree that the naturally occurring gases that warm the Earth are increasing. The lifestyles enjoyed by people in industrialized nations are causing the increase in greehouse gases.

Global temperatures have increased since the late 1800s by 0.6 to 1.2 degrees Fahrenheit. Because warmer temperatures are causing snow cover and floating ice in the Arctic Ocean to melt, sea levels have risen 4 to 10

inches over the last century. Many nations are worried that global warming might hurt their forests, crop yields, and water supplies.

It is difficult to predict exactly how much temperatures might rise before we can bring the man-made gases that cause warming under control. Most nations agree that the time to take action is now.

1. The glass roof and sides of a greenhouse allow _____ to enter and also help to keep _____ from escaping.

2. In some ways the Earth is like a greenhouse, but instead of a glass roof, our planet has naturally occurring gases like _____ _____, and _____ _____ that help to keep heat from escaping.

3. What would happen to human life without the natural greenhouse warming effect?

4. What do most scientists believe has caused recent increases in naturally occurring greenhouse gases like carbon dioxide?

    a. an increase in the sun's temperature

    b. the use of cars and other activities in industrialized nations

    c. an increase in the amount of floating ice in the Arctic Ocean

5. The story suggests that global warming might cause problems with

    a. living things that need specific temperatures to survive.

    b. radio and television reception around the world.

    c. the temperature on the surface of the moon, as well as the Earth.

6. Many nations agree that the time to take action is now. Why do you think they might feel this way?

    a. Waiting until later will increase the chances of success.

    b. Waiting until later will let the problem go away by itself.

    c. Waiting until later will increase the difficulty of the task.

Drinking Seawater

How is salt removed from seawater?

People, animals, and plants all need fresh water to live. But three quarters of the Earth's surface is covered by salt water. How do places that don't have fresh water turn salt water into fresh? The answer is *desalination*. This is a process by which salt is removed from seawater.

There are several ways in which salt can be removed from seawater. But over 60 percent of the worlds' desalinated water is produced with heat.

In desalination plants water is heated to the boiling point in large vats. This makes the water evaporate, leaving the salt behind in the bottom of the vat. The desalinated water vapor is then condensed to form fresh water in another vat.

The process of desalination is costly. But it has allowed countries in the Middle East and North Africa to develop

farming. Crops are now grown on land where they never grew before. This is also true in other parts of the world where there is very little fresh water.

Several towns in California desalinate sea-water to use for drinking. As the number of people in the world increases, more and more places will turn to desalination to provide fresh water for drinking.

1. The process by which seawater is made fresh is called _____.

2. Over 60 percent of the world's desalinated water is produced with _____.

3. When water is boiled, it becomes

    a. salt.

    b. a vapor.

    c. costly.

4. Desalinated water vapor is condensed into

    a. salt.

    b. a vapor.

    c. fresh water.

5. Desalination is costly, but it allows countries where there is very little fresh water to

    a. build large cities.

    b. farm their land.

    c. build desalination plants.

6. The story tells you that several towns in California use desalination to

    a. farm their land.

    b. provide drinking water.

    c. build houses.

The Great Sperm Whale

The numbers of sperm whales are increasing.

The sperm whale is a deep-water *mammal* (măm'əl). A mammal belongs to the group of animals that have fur or hair on their bodies. The female of the group produces milk for her babies. A sperm whale can be 35 to 60 feet long. The head of the sperm whale takes up about one-third of its huge body. Inside its mouth, there are about 60 strong, pointed teeth, each as long as a ruler.

Sperm whales eat octopus, fish, and squid. They eat about 1 ton of food a day. The whale can dive more than 3,300 feet. The average dive is between 20 to 50 minutes long.

In 1970 the Whaling Commission decided to prohibit the hunting of the sperm whale. The animal was placed on the endangered species list. Since the 1970s, the number of sperm whales has increased. In 2000, the number of sperm whales was between 500,000 to 2,000,000.

Sperm whales have a special oil in their heads. It is

used in lamps, candles, makeup, and medicines. These whales also form a gray or blackish fatty material called *ambergris* (ăm'bər grĭs') in their intestines. Ambergris is found floating on the ocean or washed up on the shore. It is used to make perfumes.

1. Animals that have fur or hair on their bodies and whose females produce milk for their young are known as _____.

2. What part of a sperm whale makes up about one-third of its body?

3. The sperm whale can be as long as _____ feet.

4. A sperm whale can dive to depths of

   a.   300 feet.

   b.   3,000 feet.

   c.   30,000 feet.

5. Why do you think scientists feared that the great sperm whale might become extinct?

   a.   The fatty material formed in the sperm whale's intestines is dangerous to the whale.

   b.   Whalers were hunting and killing the sperm whale.

   c.   The dives made by the sperm whale are too deep.

6. Under which heading would you find the best information on the sperm whale?

   a.   Mammals of the Sea

   b.   Diving for Squid

   c.   Sea Animals with Pointed Teeth

Hydrogen, Fuel of the Future

Imagine a car running on a fuel that can be made from seawater!

Such a power source may not be too far in the future. The fuel is called *hydrogen* (hī′drəjən). Hydrogen, a colorless gas, is one of the basic building blocks of which all things are made. All water contains some hydrogen.

But why would hydrogen make such a good fuel? First, it causes very little pollution. When hydrogen burns, water is released into the air and eventually returns to the Earth as rain. Also, by weight, hydrogen is more powerful than any other fuel. It can be used in every way that oil, coal, and gasoline are now used. Finally, it could be carried easily and cheaply by pipeline right to our communities.

Although hydrogen is quite plentiful, it must still be prepared for use in cars. The first step is to take hydrogen out of the water. Scientists are still trying to find ways to do this cheaply. Once hydrogen has been separated out, it must then be stored in special containers to keep it from exploding. And since hydrogen fuel takes up more room than gasoline, the fuel tanks in cars would have to be changed. Or drivers would have to refuel more often.

Perhaps someday soon, you will drive into a service station and say: "Fill it up with hydrogen."

1. Hydrogen, a basic part of all things, is a colorless _____.

2. When hydrogen burns, it gives off

    a.   oil.

    b.   smoke.

    c.   water.

3. According to the story, scientists have not yet found a cheap way to

    a.   carry hydrogen to communities.

    b.   separate hydrogen from water.

    c.   store hydrogen in tanks.

4. Hydrogen would make a cleaner fuel than gasoline.

    a.   True

    b.   False

    c.   The story does not say.

5. The story suggests that if hydrogen is used in cars, then automakers
    must design

    a.   lighter engines.

    b.   larger fuel tanks.

    c.   shorter car bodies.

6. If it is used successfully to run our cars, then hydrogen might also be used to

    a.   heat our homes.

    b.   power a windmill.

    c.   clean our water.

Putting the Sun to Work

Buildings use solar collectors to heat.

How do some people in New York City get their hot water for cleaning clothes and taking showers? They use energy from the sun, or *solar* (sō′lər) energy. Several years ago, they decided to have 200 solar collectors put on the roof of the apartment building.

The rooftop solar collectors preheat water from 55 °F to 80 °F for household uses.

A solar collector collects, traps, and absorbs (or soaks up) the sun's energy. The outer part of each collector can be made up of one or two layers of glass. The sunlight shines through the glass to the inner part. There, the rays strike a black *absorber* (ăb sôr′bər) plate. The absorber plate soaks up the sun's solar energy and changes it into heat energy. The glass layer now acts as a trap, holding in this heat energy.

Rows of tubes are attached to the absorber plate. The tubes are usually made of a metal such as copper, which is a good conductor of heat. Water flows through the tubes and is heated. Then the heated water flows to an insulated storage tank, which keeps the water hot. From there, pipes carry the hot water to kitchen sinks and bathrooms as it is needed.

Of course, solar energy can be used in other ways and in other places. Solar collectors are used in large numbers for heating outdoor pools in cool weather.

1. The word in the story that describes how a solar collector soaks up the sun's energy is _____.

2. In this story, the people used solar energy to

    a.   heat the building.

    b.   heat the water.

    c.   light the building.

3. What do the pipes do in the solar heating system described in the story?

    a.   They collect the sun's heat.

    b.   They store the heat energy collected.

    c.   They carry the heated water.

4. This story tells mainly about

    a.   how solar energy can save money.

    b.   where solar collectors should be placed.

    c.   how solar energy was put to practical use.

5. Just before it hits the absorber plate, sunlight must pass through _____.

6. A Thermos bottle can be used to keep coffee hot. The Thermos works on the same principle used in solar heating when

    a.   placing insulation around a storage tank.

    b.   putting in a back-up system of heat.

    c.   placing solar collectors on a roof.

Jane Goodall and Her Work with Chimpanzees

A well-known scientist is helping to protect chimps.

Jane Goodall is a leading research scientist on chimpanzees. She lived in the Gombe Stream Research Center in Tanzania, Africa. Dr. Goodall studied the habits of chimpanzees for more than 40 years. The more she learned about them, the better she was able to provide protection for them.

Chimpanzees are dark-haired apes native to Africa. They belong to the order of primates. This order includes human beings, marmosets, lemurs, apes, and monkeys. In the Gombe Stream Reserve area, Goddall studied chimpanzees who spent most of their time living in trees and gathering food, mostly fruit.

While studying the chimps, Dr. Goodall discovered that the chimpanzee is among the animals that use tools. Dr. Goodall observed the chimps making tools from twigs and grasses. The chimp removes the leaves from the twig with its hands or teeth. Then the chimp places the twig inside a

termite mound. When a termite bites on the twig, the chimp removes the twig and eats the termite. When the twig gets too worn out to use, the animal searches for another twig. Chimps also use twigs to clean their bodies.

After living in the reserve, Jane Goodall spent much of her time traveling around the world speaking to school groups about Roots and Shoots, an environmental program for children. In 1988 she wrote *My Life with the Chimpanzees*, her first book for young people. From this story, Jane hoped her readers would get a better understanding of the wonder of animals.

1. Jane Goodall is a research scientist who studied

    a.   lemurs.

    b.   chimpanzees.

    c.   marmosets.

2. Chimpanzees are native to

    a.   South America.

    b.   Africa.

    c.   North America.

3. Chimpanzees belong to the order of primates that include

    a.   apes, monkeys, and human beings.

    b.   apes, cheetahs, and human beings.

    c.   apes, lions, and human beings.

4. Chimps spend most of their time living

    a.   on the ground.

    b.   in trees.

    c.   in caves.

5. One of Goodall's observation included chimps

    a.   making tools.

    b.   playing in groups.

    c.   eating twigs.

6. Jane Goodall wrote about chimpanzees so readers would

    a.   get a better understanding of animals.

    b.   understand more about how chimps use tools.

    c.   learn more about her life in England.

Plants Give a Warning

It is five o'clock on a Friday afternoon. People driving their cars home from work crowd the highway. As the cars speed along, gases are pouring out from their tailpipes. The gases are left over from the burning of gasoline in a car's engine. These gases mix with the air in our atmosphere. They are called *air pollutants* (pə looͬt n'ts). Air pollutants dirty the air, making it harmful to plants and animals.

Not only cars give off pollutants. Air pollutants come out of the chimneys of many homes and factories. In fact, the burning of most fuels makes the air dirty.

Since air pollution can be harmful to us, scientists called *meteorologists* (mē tē ə rŏl' ə jĭsts) study the weather and are constantly testing the quality of the air. They use many special instruments, but there is also a simple way to detect pollution. A careful look at certain plants will often provide the first clue that the air is polluted.

In California, the death of a large forest of pine trees led to the discovery that the air in Los Angeles was polluted. At first, scientists were puzzled. Then they studied the water, the soil, and the air that surrounded the forest and found that air pollution had caused the trees to die.

The scientists found that the exhaust from cars was the main cause of the air pollution in Los Angeles.

1. Air pollutants _____ the air and make it _____ to living things.

2. Scientists use many special _____ to test air quality.

3. Air pollutants are given off when most fuels are burned.

    a. True

    b. False

    c. The story does not say.

4. The story does not tell us

    a. how plants helped scientists detect air pollution.

    b. what caused the pollution in Los Angeles.

    c. what can be done to get rid of air pollution.

5. What was the *first* clue that the air in Los Angeles was polluted?

    a. The air smelled bad.

    b. Pine trees died.

    c. People got sick.

6. Which of the following would be most likely to cause air pollution?

    a. a traffic jam

    b. a campfire in a forest

    c. a can stuffed with trash

Watching Out for Salamanders

Scientists are keeping an eye on salamanders so the animals won't become extinct.

Scientists are watching out for salamanders. Salamanders are *amphibians* that live in grasslands, wooded areas, and forests. Salamanders and other amphibians are disappearing in large numbers in some parts of Australia, western United States and Canada, Central America, and South America. Scientists believe that pesticides and other pollution are causing the salamanders to disappear.

People often confuse salamanders and lizards. Lizards have dry, scaly skins and salamanders have wet, smooth skins. Salamanders breathe through their skin and store fat in their tails. Also, lizards have claws and salamanders do not. The smallest salamander is about 2 to 5 inches long. The largest kinds are about 8 to 12 inches long.

Unlike lizards, most salamanders spend part of their lives in water. Most salamanders lay their eggs when temperatures are warm and heavy rains fall. The salamander lays its eggs in water.

Because rain washes away many pollutants out of the air, the pollutants settle on the ground or water. Sometimes the pollutants are strong enough to prevent salamander eggs from hatching. The young that do hatch cannot live in polluted water.

1. Scientists believe that salamanders are disappearing in some areas because of

     a. cold weather.

     b. pesticides.

     c. too much rainfall.

2. Salamanders are different from lizards in that salamanders

     have _____ skins.

3. The smallest kind of salamander is about _____ to _____ inches long.

4. How does pollution cause problems for salamanders?

     a. Salamander eggs crack under the heavy rains.

     b. The rains pollute the waters they lay eggs in.

     c. Salamanders cannot move during the rainy season.

5. Based on what you read, what season of the year can be most harmful to

     young salamanders?

     a. winter

     b. spring

     c. fall

6. The story does not tell us

     a. that salamanders breathe through their skin.

     b. what can be done to cut down on pesticides.

     c. the sizes of salamanders.

The Arctic Is Warming Up

The Arctic is a region around the North Pole. The region includes the Arctic Ocean, which is covered mostly by drifting ice. The Arctic is also surrounded by the most northern land areas of Canada, Russia, Finland, Sweden, Norway, Alaska, and Greenland.

There are many kinds of plants and animals in the Arctic. Some animals include musk-oxen, wolves, caribou, polar bears, and peregrine falcons. Arctic plants include mosses, lichens, grasses, shrubs, and trees.

The Arctic winter is usually long and harsh. But the weather is changing. For the last several years, native people and scientists have observed signs of warmer temperatures everywhere. They have noticed changes in ocean ice. Large stretches of ocean ice are shrinking, and the layers of ice are getting thinner and thinner. About 30 years ago, Arctic Ocean ice was 10 feet thick. Now it is about 6 feet thick.

Arctic glaciers have also melted down and shrunk over the last ten years. For example, one glacier in Alaska has shrunk by 6 miles in length. It lost about 52 square miles during the last one hundred years.

Environmentalists are very concerned about the change in the Arctic's weather. Will the

warmer temperatures affect the habitats of Arctic animals and plants? Will a longer growing season attract non-native plants and animals to the Arctic? No one knows for sure. To find answers, scientists are studying what is causing the warmer temperatures in the Arctic.

1. In the last paragraph, the word _____ refers to the place where organisms live.

2. The Arctic region is located at the

   a. North Pole and has few animals and plants.

   b. South Pole and has many plants and animals.

   c. North Pole and has many plants and animals.

3. What have the native people noticed over the years?

   a. Layers of ocean ice are getting thicker.

   b. Layers of ocean ice are getting thinner.

   c. Layers of ocean ice have not changed.

4. One Arctic glacier has shrunk by

   a. 1 mile.

   b. 3 miles.

   c. 6 miles.

5. The *main idea* of the story is that the Arctic region

   a. has many plants and animals.

   b. is getting colder.

   c. is getting warmer.

6. Which two of the following events may happen if the Arctic continues to get warmer?

   a. The length of time for hibernation of animals will change.

   b. More glaciers will be formed.

   c. Plants will have a longer growing season.

   d. The layers of ice will get thicker.

Volcanoes, a Natural Disaster

Scientists are studying volcanoes to predict when they will erupt.

In March 2000, Mount Usu volcano erupted in Abuta, Japan. A mix of smoke, melted rock, gas, and ash billowed into the air. A carpet of ash coated cars and houses. About 15,000 people had to flee their homes.

A volcano is an opening in Earth's surface through which *molten* rock and gases are forced to Earth's surface. Today, there are about 600 active volcanoes around the world. There are active volcanoes in Italy, Japan, United States, and Mexico.

Volcanologists are scientists who study volcanoes. Volcanoes are some of the most awesome natural disasters. A natural disaster can kill and injure people and other organisms, destroy buildings, and damage the environment. One of the goals of scientists is to try to establish early-warning systems. They would like to warn people to leave their homes before a volcano erupts.

One study includes using special instruments aboard Earth-orbiting weather satellites. When a crack in a volcano begins to open up, the instruments relay the information to scientists at ground stations. By studying the cracks, scientists can estimate when a volcano is about to erupt.

In 1999 scientists were able to predict the eruption of a volcano in Guatemala. A week later the volcano erupted.

1. The word *molten* means_____

2. Today there are about
    a. 300 active volcanoes.
    b. 600 active volcanoes.
    c. 900 active volcanoes.

3. Scientists can forecast when volcanoes are likely to erupt by using
    a. satellites.
    b. airplanes.
    c. rockets.

4. The purpose of forecasting volcanoes is to _____.

Use the table to answer questions 5 and 6.

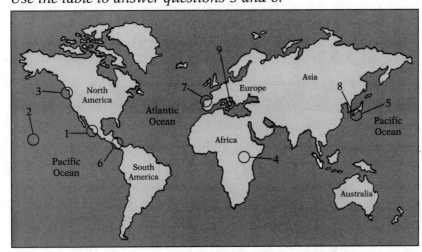

1. Colima, Mexico

2. Mauna Loa, Hawaii

3. Mount Rainier (near Seattle, Washington)

4. Nyiragongo, Zaire

5. Sakura-Jima, Japan

6. Santa Maria, Guatemala

7. Teide, Spain

8. Unzen Japan

9. Vesuvius, Italy

5. According to the map, most active volcanoes are located in
    a. France.
    b. Japan.
    c. England.

6. Which state in the United States does not have an active volcano?
    a. Hawaii
    b. Texas
    c. Washington

BIBLIOGRAPHY

Books on Life Science

_____. Human Body. New York: Time-Life, 1992.

Borland, Hal. The Golden Circle: A Book of Months. New York: Crowell, 1977.

Bright, Michael. Pollution & Wildlife, Survival Series. New York: Gloucester Press, 1987.

Casselli, Giovanni. The Human Boday and How It Works. New York: Putnam, 1987.

Cochrane, Jennifer. Land Energy, Project Ecology Series. New York: Bookwright Press, 1987.

Florian, Douglass. Discovering Trees. New York: Scribners, 1986.

Hemsley, William. Jellyfish to Insects. New York: Gloucester Press, 1991.

Henderson, Douglass. Dinosaur Tree. New York: Simon & Schuster, 1994.

Lorimer, Lawrence T.; Fowler, Keith. The Human Body: A Fascinating See-Through View of How Bodies Work, Pleasantville, NY: Readers Digest 1999

McLaughlin, Molly. Earthworms, Dirt, & Rotten Leaves: An Exploration in Ecoloygy, New York: Antheneum, 1986.

Owen, Jennifer. Insect Life. Tulsa, OK: : EDC, 1985.

Rivers, Lynn; McDonald, Sharon. Pigs, Plants & Other Biological Wonders. Dubuque, IA: Kendall/ Hunt Publishing, 1999.

Showers, Paul. How Many Teeth?. New York: Harper Collins, 1991.

Silverstein, Alvin; Silverstein, Virginia; Silverstein, Laura. Common Colds (My Health). New York: F. Watts, 1999.

Stockley, Corinne. Animal Behavior, Science and Nature Series. Tulsa, OK: EDC Publications, 1992.

Waldbauer, Gilbert. Millionsof Monarchs, Bunches of Beetles: How Bugs Find Strength in Numbers. Cambridge, MA: Harvard Univ. Press, 2000

Books on Earth-Space Science

_____. Planet Earth. New York: Time-Life, 1992.

_____. Space Planets, New York: Time-Life, 1992.

Blair, Carvel. Exploring the Sea: Oceanography Today. New York: Random House, 1986.

Bramwell, Martyn. Planet Earth. New York: F. Watts, 1987.

Bramwell, Martyn. The Oceans. New York: F. Watts, 1994.

Bramwell, Martyn. Volcanoes and Earthquakes. New York: F. Watts, 1994.

Branley, Franklin M. Sunshine Makes the Seasons. rev. ed. New York: Crowell, 1986.

Branley, Franklin M. The Sun, Our Nearest Star. New York: Crowell, 1988.

Branley, Franklin Mansfeild. Shooting Stars. New York: Crowell, 1989.

Branley, Franklin. Floating in Space. New York: Harper Collins, 1998.

Kahl, Jonathan D. W. Storm Warning: Tornadoes and Hurricanes (How's The Weather?). Minneapolis, MN: Learner Publications, 1993.

Lampton, Christopher F. Volcano. Connecticut: Millbrook Press, 1991.

Oldershaw, Cally. 3D Eyewitness: Rocks & Minerals. Dorling Kindersley, 1999.

Pollard, Michael. Air, Water, Weather. New York: Facts on File, 1987.

Selsam, Millicent E. and Joyce Hunt. A First Look at Owls, Eagles, & Other Hunters of the Sky, A First Look at—Series. New York: Walker, 1986.

Web-sites on Earth-Space Science

Hurricanes/National Severe Storms Laboratory
http://www.nssl.noaa.gov

Books on Physical Science

_____. Physical Forces. New York: Time-Life, 1992.

Barrett, Norman. Sports Machines. New York: F. Watts, 1994.

Berger, Melvin. Atoms, Molecules, & Quarks. New York: Putnam, 1986.

Bramwell, Martyn. Glaciers and Ice Caps. New York: F. Watts, 1994.

Branley, Franklin M. Gravity Is a Mystery. rev. ed. New York: Crowell, 1986.

Branley, Franklin Mansfield. The Beginning of the Earth. New York: Harper Row, 1988.

Haslam, Andrew; Glover, David. Machines (Make It Work! Series). Pittsfield, MA: World Book, Inc., 1997.

Morgan, Nina. Lasers (20th Century Inventions). Texas: Raintree Steck-Vaughn, 1997.

Walpole, Brenda. Movement, Fun With Science Series. New York: Gloucester Press, 1987.

Whyman, Kathryn. Heat and Energy. New York: Gloucester Press, 1986.

BIBLIOGRAPHY

Books on Environmental Science

Accorsi, William. Rachel Carson, New York, NY: Holiday House, 1993.

Bonnet, Robert L. Environmental Science. Tab Books, 1990.

Forman, Michael H. Arctic Tundra (Habitats). Danbury, CT: Children's Press, 1997.

George, Jean Craighead. 1 Day in the Tropical Rain Forest (Newbery Medal Winner Series, No. 5). New York: Crowell Co., 1990.

Miller. Race to Save the Planet. Pacific Grove, CA: Brooks/Cole Publishing Company, 2001.

Mongillo, John; Zierdt-Warshaw, Linda. The Encyclopedia of Environmental Science. Phoenix, AZ: Oryx Press, 2001

Morgan, Sally. Acid Rain (Earth Watch). New York: F. Watts, 1999.

Pollock, Steve. The Atlas of Endangered Animals (Environmental Atlas Series). New York: Checkmark Books, 1993.

Pringle, Laurence P. Rain of Troubles. New York: Macmillan; Collier Macmillan, 1988.

Ripple, Jeff. Manatees and Dugongs of the World. Stillwater, MN: Voyageur Press, 1999.

RECORD KEEPING

The Progress Charts on these pages are for use with questions that follow the stories in the Life Science, Earth-Space Science, and Physical Science Units. Keeping a record of your progress will help you see how well you are doing and where you need to improve. Use the charts in the following way:

After you have checked your answers, look at the first column, headed "Questions Page." Read down the column until you find the row with the page number of the questions you have completed. Put an X through the number of each question in the row that you have answered correctly. Add the number of correct answers, and write your total score in the last column in that row.

After you have done the questions for several stories, check to see which questions you answered correctly. Which ones were incorrect? Is there a pattern? For example, you may find that you have answered most of the literal comprehension questions correctly but that you are having difficulty answering the applied comprehension questions. If so, then it is an area in which you need help.

When you have completed all of the stories in an unit, write the total number of correct answers at the bottom of each column.

PROGRESS CHART FOR EARTH-SPACE SCIENCE UNIT

Questions Page	Comprehension Question Numbers				Total Number Correct per Story
	Science Vocabulary	Literal	Interpretive	Applied	
43	1	2,3	4	5,6	
45	1	2,3	4,5	6	
47	1	2,3	4,5	6	
49	1	2,3	4,5	6	
51	1,2	3,4,5	6		
53	1	2,3	4,5	6	
55	1	2,3,4	5,6		
57	1	2,3	4,5	6	
59	1	2,3,4		5,6	
61		2,3	4,5	1,6	
63	1	2,3	4,5	6	
65	1	2,3	4,5	6	
Total Correct by Question Type					

PROGRESS CHART FOR LIFE SCIENCE UNIT

Questions Page	Comprehension Question Numbers				Total Number Correct per Story
	Science Vocabulary	Literal	Interpretive	Applied	
7	1	2,3	4,5	6	
9	1	2,3	4,5	6	
11	1	2,3	4,5	6	
13	1	2,3	4,5	6	
15	1	2,3	4,5	6	
17	1	2,3	4,5	6	
19	1	2,3	4,5	6	
21	1	2,3	4,5	6	
23	1	2,3	4,5	6	
25	1	2,3	4,5	6	
27	1,2	3	4	5,6	
29	1	2,3	4,5	6	
31	1	2,3	4,5,6		
33	1	2,3	4,5	6	
35	1	2,3	4,5	6	
37	1	2,3	4,5	6	
Total Correct by Question Type					

PROGRESS CHART FOR PHYSICAL SCIENCE UNIT

Questions Page	Comprehension Question Numbers				Total Number Correct per Story
	Science Vocabulary	Literal	Interpretive	Applied	
71	1	2,3	4,5	6	
73	1	2,3	4,5	6	
75	1	2,3	4,5		
77		1,2,3,4	5	6	
79	1	2	3,4	5,6	
81	1,2	3	4,5	6	
83	1	2,3,4,5	6	5,6	
85	1	2	3,4	5,6	
87	1	2,3	4,5		
89	1	2,3,4,5	6		

Total Correct by Question Type

PROGRESS CHART FOR ENVIRONMENTAL SCIENCE UNIT

Questions Page	Comprehension Question Numbers				Total Number Correct per Story
	Science Vocabulary	Literal	Interpretive	Applied	
95	1	1,2	3,4,5	6	
97	1	2,3,4,5,6			
99	1	2,3	4,5	6	
101	1	2,3	4,5	6	
103	1	2,3	4,5	6	
105	1	1,2,3,4,5	6		
107	1	2,3	4,5	6	
109	1	1,2,3,4	5,6		
111	1	2,3,4	5,6		
113	1	2,3	4	5,6	

Total Correct by Question Type

METRIC TABLES

This table tells you how to change customary units of measure to metric units of measure. The answers you get will not be exact.

LENGTH

Symbol	When You Know	Multiply by	To Find	Symbol
in	inches	2.5	centimeters	cm
ft	feet	30	centimeters	cm
yd	yards	0.9	meters	m
mi	miles	1.6	kilometers	km

AREA

Symbol	When You Know	Multiply by	To Find	Symbol
in^2	square inches	6.5	square centimeters	cm^2
ft^2	square feet	0.09	square centimeters	cm^2
yd^2	square yards	0.8	square meters	m^2
mi^2	square miles	2.6	square kilometers	km^2
	acres	0.4	hectares	ha

MASS (WEIGHT)

Symbol	When You Know	Multiply by	To Find	Symbol
oz	ounces	28	grams	g
lb	pounds	0.45	kilograms	kg
	short tons (200 lb)	0.9	tonnes	t

VOLUME

Symbol	When You Know	Multiply by	To Find	Symbol
tsp	teaspoons	5	milliliters	mL
Tbsp	tablespoons	15	milliliters	mL
fl oz	fluid ounces	30	milliliters	mL
c	cups	0.24	liters	L
pt	pints	0.47	liters	L
qt	quarts	0.95	liters	L
gal	gallons	3.8	liters	L
ft^3	cubic feet	0.03	cubic meters	m^3
yd^3	cubic yards	0.76	cubic meters	m^3

TEMPERATURE (exact)

Symbol	When You Know	Multiply by	To Find	Symbol
°F	Fahrenheit temperature	5/9 (after subtracting 32)	Celsius temperature	°C

METRIC TABLES

This table tells you how to change metric units of measure to customary units of measure. The answers you get will not be exact.

LENGTH

Symbol	When You Know	Multiply by	To Find	Symbol
mm	millimeters	0.04	inches	in
cm	centimeters	0.4	inches	in
m	meters	3.3	feet	ft
m	meters	1.1	yards	yd
km	kilometers	0.6	miles	mi

AREA

Symbol	When You Know	Multiply by	To Find	Symbol
cm^2	square centimeters	0.16	square inches	in^2
m^2	square meters	1.2	square yards	yd^2
km^2	square kilometers	0.4	square miles	mi^2
ha	hectares (10,000 m^2)	2.5	acres	

MASS (WEIGHT)

Symbol	When You Know	Multiply by	To Find	Symbol
g	grams	0.035	ounces	oz
kg	kilograms	2.2	pounds	lb
t	tonnes (1000 kg)	1.1	short tons	

VOLUME

Symbol	When You Know	Multiply by	To Find	Symbol
mL	milliliters	0.03	fluid ounces	fl oz
L	liters	2.1	pints	pt
L	liters	1.06	quarts	qt
L	liters	0.26	gallons	gal
m^3	cubic meters	35	cubic feet	ft^3
m^3	cubic meters	1.3	cubic yards	yd^3

TEMPERATURE (exact)

Symbol	When You Know	Multiply by	To Find	Symbol
°C	Celsius temperature	9/5 (then add 32)	Fahrenheit temperature	°F